Honey in the Comb

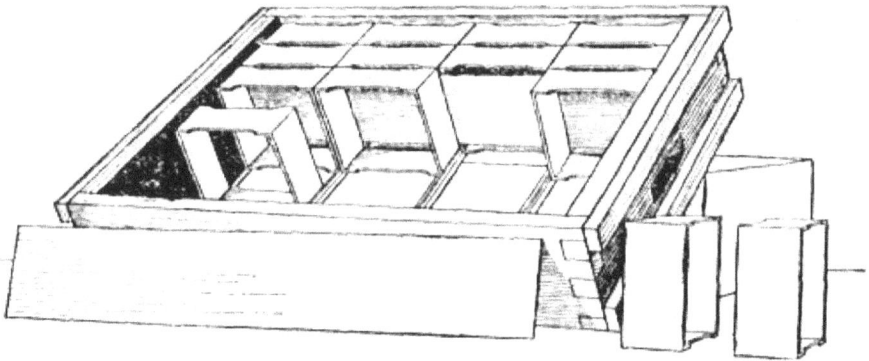

by

Carl E. Killion

Honey in the Comb
by Carl E. Killion

Original publication 1951
Reprint © 2014

Cover background photo courtesy of Walden Ridge
Honey, Sale Creek TN

X-Star Publishing Company
Nehawka, Nebraska, USA
XStarPublishing.com

ISBN 978-1-61476-101-3

Pages 158
Illustrations 72

Author's Dedication

This book is dedicated to my good wife Elizabeth who has been my constant helper in my bee work through-out all the years; to my son Eugene who has chosen comb honey production as his life's work; and my son Carl who, with his camera, has been most helpful in the making of this book.

Preface

The increased demand for section comb honey, or honey in the comb, is the incentive which has encouraged the writing of this book.

The production of fancy comb honey has for many years been far below the demand for this fine article.

Many times the writer will mention the names of Dr. C.C. Miller and Charles A. Kruse. I am indebted to both of these men for the valuable contributions they have made. I have perhaps followed the teachings and general practices of Dr. C.C. Miller more than any living person. Although I never met this fine man, I feel as though I knew him quite well.

This book is a result of a lifetime study of the equipment and practices for the production of fine comb honey. It is also the purpose of this book to encourage only the finest quality of comb honey to be offered to our markets.

The writer does not insist to the reader that success is only possible when using all the special equipment as used by him. The only exception is that I am convinced that the ventilated "T" super is the most valuable piece of equipment for comb honey, besides the one-piece honey section.

As this is being written I am saddened by the fact that men have been in Washington asking for price support for extracted honey. We who keep bees must not believe we have some divine purpose here on earth but are just working with man's most useful and interesting insect. We have, however, lost our dignity when we seek price support for such a fine product as honey. It will be a sad day indeed when we must seek support for comb honey. Comb honey must always keep its place in the sun and sell itself as it has always done.

The author desires to express his appreciation to Minnie S. King of the American Bee Journal for her critical assistance in the preparation of the manuscript of this book.

Carl E. Killion

Transcriber's Note

I chose to republish this book because it is one of the best books on the art of raising comb honey and due to unfortunate circumstances the original books are in short supply. Eugene Killion, his son, has a book by a similar title, which is also good reading, but it is a different book in many ways. I would read both, but forced to choose between them I would consult this one. The one advantage to Eugene's book is it is more up to date as far as equipment.

Some of the equipment may be unavailable in the current market or limited to only a few suppliers. Almost all of this, however, translates well to any of the newer methods of making comb honey in other cassettes such as Ross Rounds. Even if one does not wish to raise comb honey as your primary product, raising comb honey is another of those beekeeping pursuits that teaches you much about bees and that makes it worthwhile in itself.

—Michael Bush

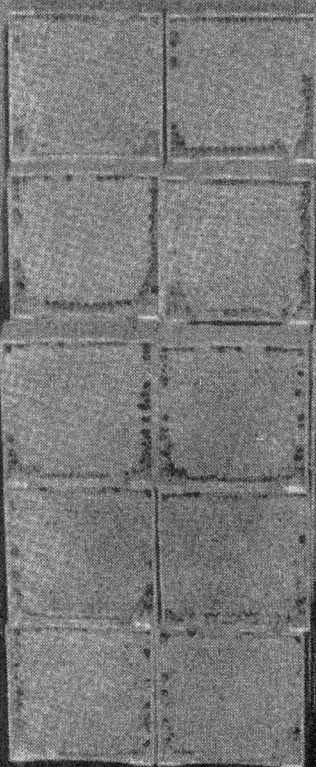

Table of Contents

Table of Figures

I. Comb Honey

COMB honey is one of the most beautiful agricultural products in the world. To me, it is the most beautiful. It takes so few words to describe it, honey in the comb tells its own story as no human could ever describe it. It is pure, delicious honey in its natural original state. It is nature's finest sweet in its own original package as the bees have made it throughout the ages. Comb honey is a product that the intelligence and ingenuity of man cannot duplicate or equal.

One might say that all honey is good, but honey in the comb is the best. One can offer honey to the public just as the bees make it, nothing added, nothing taken away. Each cell of crystal clear honey is enclosed in pure beeswax and retains all the fragrance of the fields of sweet blossoms. The delicate aroma of flowers, which is lost to a certain degree in extracting honey, may still be found in comb honey. All of these desirable features have been carefully weighed in my mind, and my wish is that the production of this wonderful product will be continued.

Comb honey must never be given up entirely for liquid honey. We must at all times have comb honey to show what God's most useful insect gives to man in the form of sweetness in its original state. It is clean and attractive in appearance, and a convenient size package for the consumer. Its delicate aroma and flavor have not been injured by heating or processing, as sometimes is the case with extracted honey.

The work must be carefully planned and executed with the most exacting precision. One cannot afford to make too many mistakes as they are expensive and our seasons are so short the mistakes cannot always be corrected in time. The honeyflows some years are of short duration; time is valuable when nectar is coming in and comb is being built.

One must always remember, "The quality or grade of comb honey is built to the degree of perfection while the super is still upon the beehive; man can improve it so little when removed from the hive."

The extracted honey packer can improve some on the quality of honey after it reaches his plant. If the liquid honey has not been properly strained or heated to prevent crystallization, he has the processing equipment to do this job. The packer may use two or more different flavors or colors of honey to blend into the certain flavor or color combination desired. Nothing like this is possible when producing comb honey.

It is logical to assume that beekeepers would want to produce more comb honey on account of the relatively poor market for liquid honey. For the past fifteen years it has required little effort to sell comb honey—it almost sells itself. The demand has been far greater than the supply.

Most men do not like to take the time and effort required to produce quality section comb honey. The cry has been more and more of this and that until man has pushed mass production to the highest degree. The extracted honey producer has been caught in this mad whirlwind until now there are literally mountains of extracted honey in warehouses begging for a market.

Fig. 1. The only place for comb honey is on the table.

It is true that there are many more uses for extracted honey than there are for comb, but it has always been the writer's contention that the place for honey was upon the table. Liquid honey has my fullest respect, but it does not appeal to me in the same way that honey in the comb does. Man has removed the honey from the very thing that associates it with the true original product—the comb.

One cannot follow any hit-or-miss, let-alone, or neglectful practices in producing comb honey. The writer has seen some fairly good crops of extracted honey produced, and wondered how it was done after observing how careless the beekeeper performed his work. Perhaps the first thing to consider is whether one expects to be careful and exacting in the work necessary to make section comb honey a profitable enterprise. Unless one is careful and exacting, it would be well not to attempt working for comb honey.

It is the purpose of this book to assist anyone who wishes to have the finest quality section comb honey with maximum production at a minimum cost in labor and expense. One can very easily go to the extreme of making the enterprise too costly and reduce the profit to a degree where it would not pay to continue the production of section comb honey.

The writer has spent over thirty years in the careful selection of proper equipment and practices to enable him to have the finest quality possible at a moderate expense.

For many years all the bee supply companies have made equipment most suitable for extracted honey, giving little or no thought to equipment for comb honey. Some have even discouraged people from wanting comb honey equipment. We were forced, therefore, to manufacture much of our own equipment as we were unable to purchase the necessary material from the bee supply companies. I do not recommend homemade equipment over factory equipment as my experience as an apiary inspector has proved how faulty much of this homemade equipment may be. If bee supply manufacturers cannot supply the required items it is therefore necessary to have them made elsewhere.

Comb honey may be secured with an assortment of equipment, but it can be obtained much easier and a finer product realized if the equipment used is designed for its production.

The equipment needed to produce and market comb honey does not require as great an investment as that required for extracted honey. One does not require an extractor, uncapping knives, capping melter, honey storage tanks, or processing equipment.

There is one serious drawback which prevents many beekeepers from being interested in comb honey—the tendency of bees to swarm. This has perhaps discouraged more beekeepers than all other things combined. It is true that when a colony is crowded into comb honey supers they soon develop the desire for swarming. The beekeeper must then work for the control of swarming as quickly as possible. There are many control methods but not all are profitable. The writer has followed a practice of swarm control that has proved to be both efficient and profitable.

Fig. 2. An ideal windbreak for an apiary.

II. Choosing a Location

THERE are many things to consider when choosing an apiary site. Areas where there is an abundance of nectar-bearing plants to give a copious flow over a long period of time are desired. The plants should produce nectar light in color, as areas that produce dark honey or mixture of dark and light honeys are not to be desired. A mixture of colors in the comb honey decreases its sale value. If the light honeyflow is not followed too closely by one of dark honey, it is best. In this way the honey does not mix in the super. One of the writer's apiaries is located where the clover honey is light in color and of fine flavor; later in the season there is a mixed flow of darker honey from goldenrod, Spanish needle, and heartsease. The two flows are separated by at least three weeks so we do not have any mixture of the two colors.

Some of our heavily wooded areas are not good to choose, as bees will gather an excessive amount of propolis. This propolis will stain the wood and surface of the comb, which is undesirable. The amount of labor required to remove the propolis from the sections is greater.

Apiaries that are placed where they are unprotected from the wind should have an artificial windbreak. One of the best windbreaks the writer has seen is the high board fence shown in Fig. 2. The boards are spaced far enough apart to permit the wind to filter through the cracks. The Osage orange which is much higher than the fence prevents the wind from dripping over the tip of the fence and hitting the hives.

It is best to place apiaries on the side of a hill or on a sloping ground; never on hilltops or the lowest part of an area. Hilltops receive the full force of the wind and the low places do not allow proper air drainage. Air drainage is very necessary to prevent moldy combs in the winter and permits better ripening of honey in the summer.

Contrary to what some may say, we want hot nights during our honeyflows. Some claims have been made that plants produce nectar most freely when days are hot and nights are comparatively cool. The writer has not found this to be entirely true.

The comb is more tender when built during hot nights than when built during cool nights as the hot nights give the wax workers a chance to build comb more easily in all parts of the super. Hot nights prevail during June and July almost anywhere in the Mississippi Valley. In some parts of Florida and California where the nights are cool, the combs are not quite so tender as in areas where nights are hot.

Some honeys granulate quicker than others and if the producer can prevent such honeys from being stores in sections it is best to do so. Alfalfa honey is beautiful and clear but it soon granulatesand the same is true of aster honey.

Almost anyone can secure good results in producing comb honey when there is a big honeyflow. The bees just always seem to do the right thing without too much effort upon the part of the owner. These big flows do not occur every year, perhaps an average of one year in five or six. One must make every effort to produce as good a grade of comb honey as possible in years of average honeyflows. That is one advantage of having equipment which has been carefully designed for comb honey.

Outapiaries

All commercial honey producers must follow the practice of having their bees in outapiaries, and some are many miles from home. We even find some beekeepers with only a few colonies having some hives away from the home apiary. This may be an expensive thing to do unless the owner has reasons for doing it. Some outapiaries are placed in ideal locations; others just placed anywhere to get them off the truck.

The writer has quite a problem to find many ideal locations for an outapiary in our flat Illinois prairie land. Our present agricultural practices do not allow much wasteland; in fact in some areas we do not see a group of shade trees large enough to place many hives underneath. We prefer partial shade, some wind protection and a natural water supply. One reason we prefer partial shade is because we cannot stagger our supers for ventilation in hot weather. A small amount of shade prevents hives from becoming too hot and forcing the bees out of the supers. We also find it more comfortable for ourselves to have a little shade when working the bees.

Apiaries should be placed where a truck can be driven as near as possible, preferably between the rows of hives. Comb honey supers are never quite as heavy as extracting ones and may be carried a short distance. In areas where there is a danger of hives being molested and honey stolen it is best to place the apiary where the owner of the land can watch for such conditions, and report if hives are molested. The bees should be placed near enough to a natural water supply to keep them from the farmer's water tanks where the stock drink. If a natural water supply is not available the beekeeper should supply water in such a way the bees will use it instead of the stock tank. We find water is used in early spring brood rearing and in extremely hot weather. In the article "Outapiaries," by the writer in the January 1941 issue of the American Bee Journal are photos of bee yards, some showing excellent windbreaks. The writer also calls attention to adequate air drainage. For our flat terrain we cannot always find a place which gives air drainage to any extant. Proper air drainage will help reduce excess moisture in the hives. The writer believes excess moisture in the hives will kill more bees than the cold weather itself.

It is desirable to locate apiaries in disease free or reasonably disease free areas, and where nectar is expected to be most abundant, for after all it is the honey crop that counts. It has already been mentioned about producing section honey where plants produce nectar which is light in color. An area which produces a mixture of light and dark honeys at about the same season should be avoided where possible.

Honey Plants and Bee Pasture

The major honey plants in this area are not too numerous, the same might be applied to the list of minor honey plants. Most of our apiaries are located in an area where intensive farming of cash grain crops is followed. Farms of 400 to 1000 acres are quite common and most of these farms have not had the proper rotation of grain with legumes. Very little wasteland or even pasture land is to be found in some areas. The result is that beekeepers must depend upon planted crops for most of their honey. Sweet clover is the major honey plant today. Some years when white Dutch clover is plentiful it adds its share to our crop. Alsike clover which was once found in some areas is not grown any more. A field may be found now and then but not as often as fifteen or twenty years ago. Our main fall crop is heartsease and due to our modern mechanized methods of farming heartsease is not as plentiful as it was a few years ago.

Fig. 4. Sweet clover—a major source of honey throughout a large portion of the United States.

A short distance away from this flat area one may find other honey plants which may be either a major or minor source of nectar depending upon the location. East and south are locust, basswood, poplar, asters, golden-rod, and Spanish needle. The last two years considerable interest has been taken in the planting of Ladino clover. This plant should be found valuable in areas where it is grown in abundance. We hear beekeepers refer to it often as the new Ladino clover. On page 121 in Dr. C.C. Miller's book "Fifty Years Among the Bees," he mentions getting some Ladino seed from Switzerland and growing some plants in 1902. One can see how long it has taken this clover to receive attention in this country.

Fig. 5. White Dutch clover, the dean of pasture legumes, sometimes adds its share to the honey crop.

The last two seasons we have moved some of our apiaries from the flat prairie to the pasturelands and more or less forest areas south of us for a fall fill-up on aster and goldenrod. It may become necessary in the future for beekeepers to encourage the growing of Hubam for our bees to secure honey for winter feed. In this same pasture area we should make an effort to see if some of our honey plants such as bird's-foot trefoil and vetches cannot be grown at a profit to both the cattleman and the beekeeper.

Some new developments are now being tried in planting various seeds. The idea is to take sees of odd shapes and sizes and give them a coating to make them perfectly round for more accurate mechanical planting. The coating of the seed will contain fertilizer, fungicide, insecticide, and bird repellent. The whole idea is new but interesting and may prove to be of great help to honey plants in the future.

Fig. 3. The author in one of his favorite outyards.

III. Comb Honey Equipment

IN the comb honey era before World War I, most of the equipment was eight-frame size. It was at the start of the production of extracted honey that the ten-frame size hive became so popular. The writer started his bee-keeping career with mostly ten-frame hives as he only had about fifteen eight-frame hives when the change was made to ten-frame size. This choice was more by accident than by intelligent planning. Two of the greatest comb honey men, Dr. C.C. Miller and later Charles A. Kruse, used the eight –frame size exclusively. In my five years of partnership with the latter, we used both sizes of hives, Kruse favoring the eight-frame, and I the ten frame. During the last three years our records showed the ten-frame hive gained 15-16 sections over the eight frame in the three year period. At present prices of about $8 per case this would amount to approximately $1.66 more each year for the standard ten-frame hive.

The Hive

One could hardly call the hive I use a ten-frame hive; it is standard ten-frame Langstroth size in every respect, but only nine frames are used. We find any number of beekeepers using only nine frames in a ten frame hive body, but the frames are equally spaced and occupy all the space that ten frames would occupy. The nine frames used in our hive body are spaced their regular distance apart, and a follower board is used on each side next to the side wall of the hive. Many beginners have never heard of a follower board unless they run across it in some old bee literature. In our system of management the follower is a necessary part of our equipment, just as necessary as a hive cover or bottom board. The follower is used for more than one reason. The board can be removed much easier than a frame if all ten frames where used. It permits the removal of the first brood frame with ease and later returned to the hive before the next one is removed, as the space is adequate for this purpose.

One need not place a frame of brood outside the hive at any time. If both boards are removed it makes an ideal space for feeding with a division board feeder, and again one does not have to find a place to store surplus combs. In queen rearing, the cell-bar frame can be placed anywhere desired in the brood chamber after the two boards are removed, preferably as near the center as possible. The boards permit better ventilation and insulation to the brood. The main reason is we get more brood in the nine frames used than most beekeepers get in ten frames. In actual experience we find a queen does not like to lay in the two outside combs in a ten-frame hive body, next to the wall of the hive, the reason perhaps being the sudden change in temperature. The result, we find in nearly every hive, is only eight frames of brood instead of ten. One more advantage to using only nine frames spaced and arranged as they are, is that our sections in the super are directly above frames of brood. By having all sections over brood, the bees work in them better than when sections are above solid frames of honey. One will find the frames of honey on the sides if all ten frames are used. Right here it might be well to mention that this is one of the reasons we like the eight-frame size super in preference to the ten-frame size.

Fig. 6. 9 frames in the brood chamber instead of 10. A follower board is used on each side to take up space.

Fig. 7. The brood chamber with the follower boards removed and a division-board feeder placed at one side.

Fig. 8. The Miller bottom board with rack or false bottom in place.

The Bottom Board

The bottom board was copied from the one used by Dr. C.C. Miller. He describes it as "a plain box, two inches deep, open at one end." In his book, "Fifty Years Among the Bees," he writes: "A very nice thing in winter, and any other time when there is no danger of the bees building down, but quite too deep for harvest time." Later the good Doctor did try them for summer and continued to use them throughout the entire year. He had to make a slatted arrangement to be used beneath the combs which prevented the bees from building their combs down below the frames.

We are convinced that the deep bottom board is just as essential in the summer as in winter. Dr. Miller perhaps never mentioned it but it does lessen the desire to swarm where it is used. Here in our area we have such hot, humid days and nights and the large two-inch entrance gives each colony a chance to breathe, so to speak. This large entrance helps in the immediate evaporation of nectar which is so important in hives with limited storing space, such as practiced with crowding comb honey colonies.

Fig. 9. The entrance closer with winter opening. In spring, it is turned over and rests on the nail heads.

The two-inch entrance gives enough ventilation many times in moving bees that it does not require additional top or bottom screening. When moving bees in hot weather we use a top screen in addition to the one in the entrance. For many years we have been using an entrance screen made of ordinary window screen which is about twelve mesh and bound around the edges with tin. Recently we tried eight mesh hardware cloth and like it best. The wire is rather stiff and will stand considerable abuse and does not need to be bound on the edges.

The slatted arrangement for preventing the bees from building combs beneath the frames, we call a "false bottom." We leave this false bottom in the hive bottom winter and summer. The bees may like it removed in winter so strips of wood do not interfere with clustering space. We tried removing them one fall and did not replace them in time next spring. The dandelion flow as starting when we tried to replace them; the bees had already built many pieces of comb down almost two full inches. Instead of pushing the false bottoms into the entrance of the hive as they were removed the hive body had to be raised and all this comb removed.

The railings of this false bottom are 18 inches long, 15/16 inch in width and about 9/16 inch in thickness. The crosspieces are 13 7/8 inches long, ¾ inch wide, and 5/16 inch thick. The wide crosspiece in front is about 4 inches wide and its use is to prevent bees from chewing out the corners of the combs in front. All the boards are spaced approximately 5/16 inch apart. Staples are driven into the bottom of the railings to space it away from the bottom board. Nails or staples are driven into the railings at the back and the sides to make a bee space all around. The top of this false bottom should be 5/16 inch shallower than the top of the bottom board. The staples used in the railings and sides will permit the removal of the false bottom at any time. The bees do not blue it fast as is the case if wood is touching against wood.

For winter, an entrance block must be used to reduce the large entrance. This block is made of regular ¾ inch lumber and is 2 ¾ inches wide. One edge of the block is notched 5/16 x 3 ½ inches which is used as the winter entrance. The other edge has two nails driven to within 3/16 inch of the heads and are spaced about 2 inches from each end. The boards are turned over to rest on these nail heads in early spring. This gives an entrance 5/16 of an inch high and full width of the hive. Nails are used to hold the block in place against the front of the hive. The block may be raised as the colony gains strength and the weather gets warmer. The block later may be removed entirely.

The Hive Cover

Our cover was patterned after the Miller cover except the front and back of the cover extend down over the edges of the hive about one-half inch. The cover does not extend beyond the sides of the hive but is made the same width as the hive. This is a desirable feature when moving bees on trucks as less space will be wasted. In packing the bees for winter it permits the hives to be shoved together for better convenience. The cover is made with a so-called dead air space between the wooden parts. For several years now we have been using tempered presswood underneath and wood on the upper side, covered either with plain asphalt paper, tin, galvanized iron, or aluminum.

No weights are necessary to hold the cover in place on the hive in windy weather. The inner cover is not used and the bees may soon deposit enough glue on the cover to hold it in place regardless of how the wind may blow. The handling of just one cover is quite a laborsaver and timesaver in the apiary.

For three years we have been taking care of fifty colonies owned by another party. These colonies are used for orchard pollination and are all in new factory-made equipment with outer metal and inner covers. Each cover must be weighted down with at least two paving bricks. We find more time is spent in going through these fifty colonies than sixty of our own where only one cover is used.

Brood Frames

The brood frame we use was patterned after the one used by Dr. C.C. Miller and we call it the Miller frame. The frame is Langstroth size and is spaced with nails instead of the Hoffman type as are in general use today. The nails are driven into the wood until only a quarter of an inch shows. The only points of contact between the frames are the nail heads against wood. We find that our bees are less inclined to propolize the frames together when nail spaced and the bee space remains more accurate between the frames. We find the space grows larger each year when Hoffman spaced frames are used on account of the propolis. We expect less bees to be killed when pushing the frames against each other since the only points of contact are the nail heads against wood. Ventilation should be perfect around this frame. The short top bars allow a bee space at the end of the frame which minimizes propolis and permits the easy removal of frames.

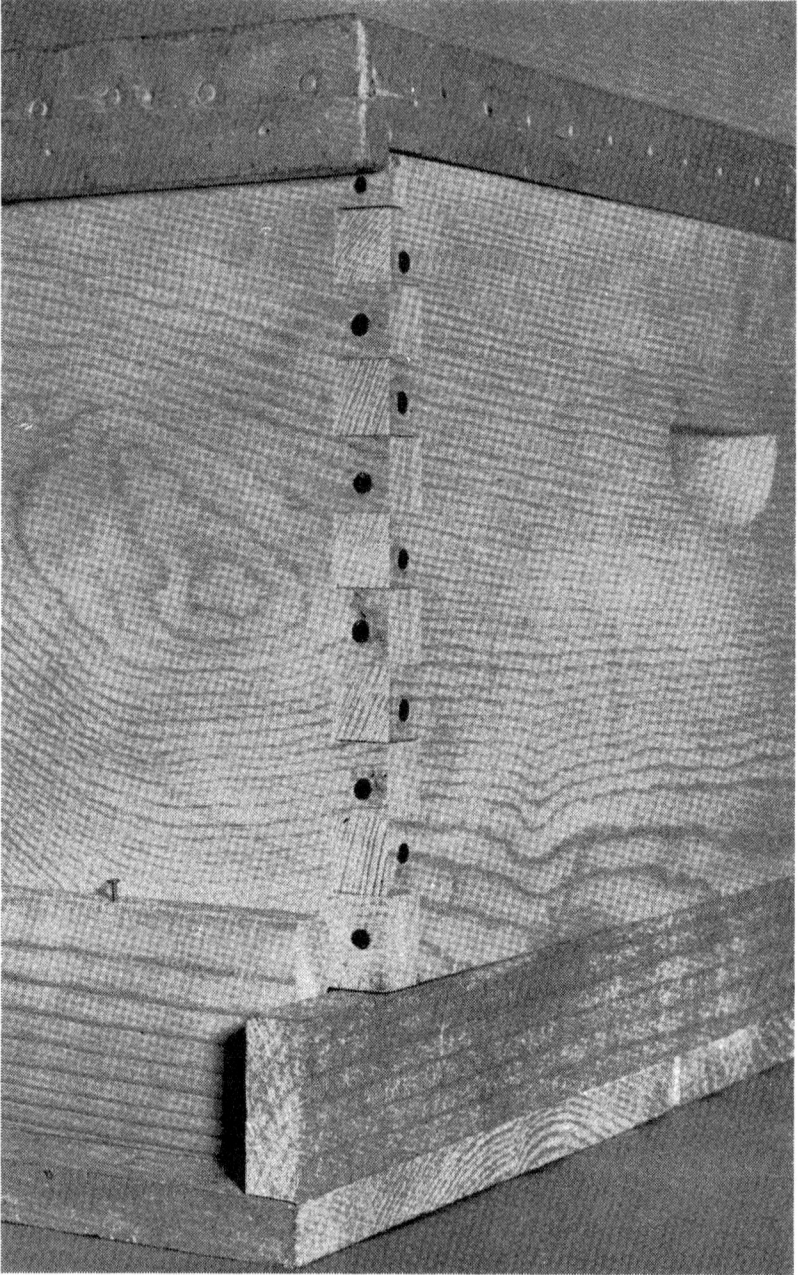

Fig. 10. Corner of the hive showing part of the cover
which does not extend beyond the side of the hive.

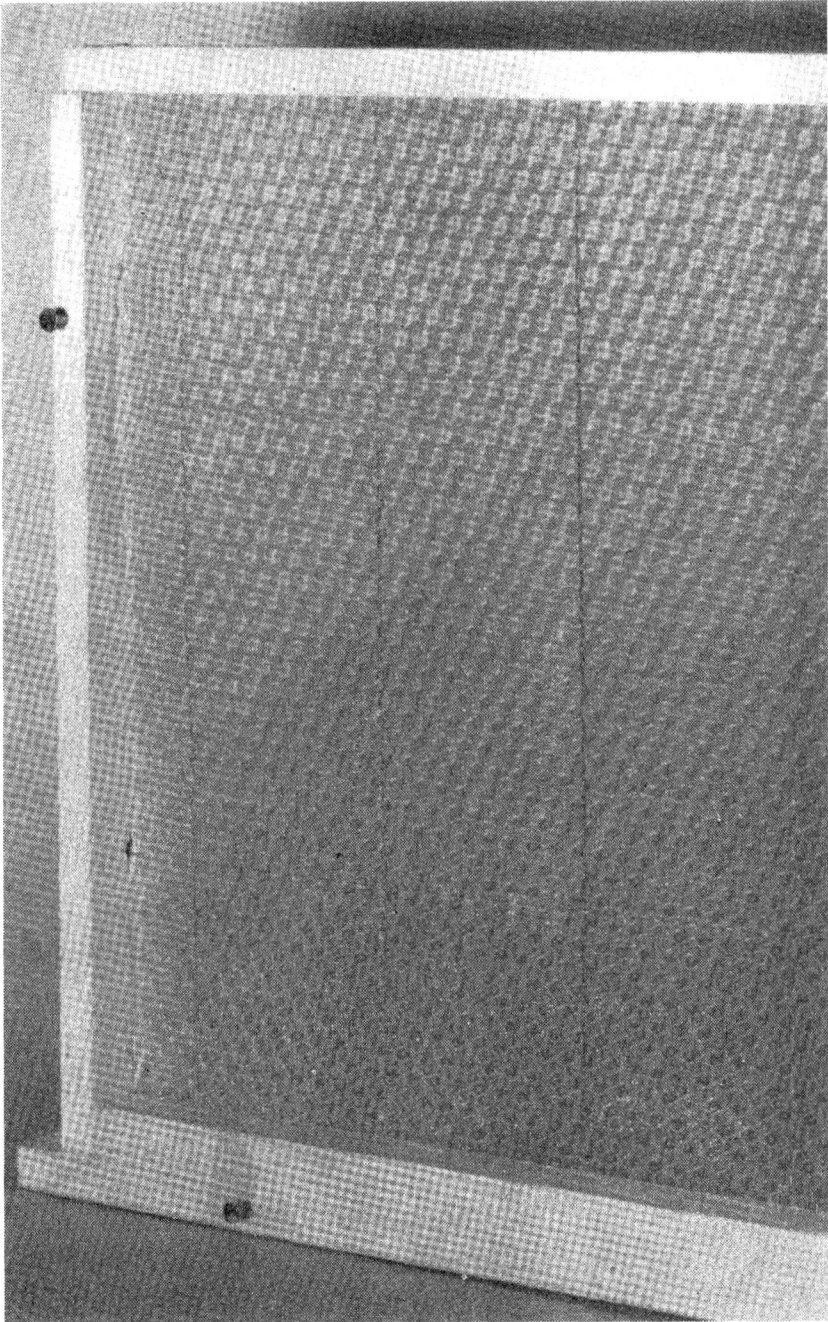

Fig. 11. End of frame showing system of nail spacing.

From time to time the writer had read articles where beekeepers were using the deep frame hive for section comb honey or were producing it over a double brood chamber. Some of the methods described in the articles would take about all of the hive bodies, bottoms and covers one bee supply house could make in a year. One can get so much equipment in use that it becomes un-profitable. I have tried the deep frame hive, eight, ten and eleven frames, also the two body system. I have elevated brood above the supers, put brood at the sides and tried about every scheme possible to improve the production and quality of the comb.

I believe Rev. Langstroth designed a much better hive than he ever realized. The beekeeper who can pro-duce excellent comb honey on a deep frame hive or over a double brood chamber is a much better beekeeper than I ever expect to be.

Our Ventilated T Super

The ventilated T super is our most prized piece of comb honey equipment, with the exception of the honey section itself. The super has not been too popular. To those who have not been familiar with this type of super I would like to give some history of it.

During the North American Beekeepers' Convention in Toronto, Canada, in 1883, D.A. Jones of Beetown, Ontario, showed the super to Dr. C.C. Miller. No mention has been made as to whether Mr. Jones invented the super or not; perhaps he did. He may have taken the idea from the Heddon super. Dr. Miller used the Heddon super in the year 1883 before he saw the T super. He wrote concerning Mr. Jones' super: "I was much im-pressed with it. The next year I put a number of T supers in use, and the more I tried them the better I liked them. I have tried a number of other kinds since, but nothing that has made me desire to make a change."

The super was called a T super because the sections rest upon tin supports which are bent to form an inverted T. Because the wooden section holders were not used, the supers were made shorter than those having the section holders. The supers were also shorter than the length of the hive body on which they were to be used. To make them long enough it was necessary to nail wooden strips on each end of the super, both top and bottom edges. These strips increased the length of the super to that of the hive body. The top strips acted as handholds for handling the super.

Dr. Miller used eight-frame equipment and his supers with the regular thickness of lumber were too narrow to permit the use of a follower on each side of the rows of sections. He had to place one row of sections against the sides of the super which resulted in this row of sections being of lower grade than the rest. His supers measured 12 1/8 inches inside width; our eight frame size is 12 ¾ inches wide inside.

The first T supers used by Charles A. Kruse were constructed from the same pattern as used by Dr. Miller. He, too, could use only one follower, but decided on making the sides of the super thinner so a follower would be used on each side. This improved the super a great deal. Kruse and I later added ventilation to the ends of the super. This additional ventilation with a bee space completely around the group of sections has made the super complete in every detail.

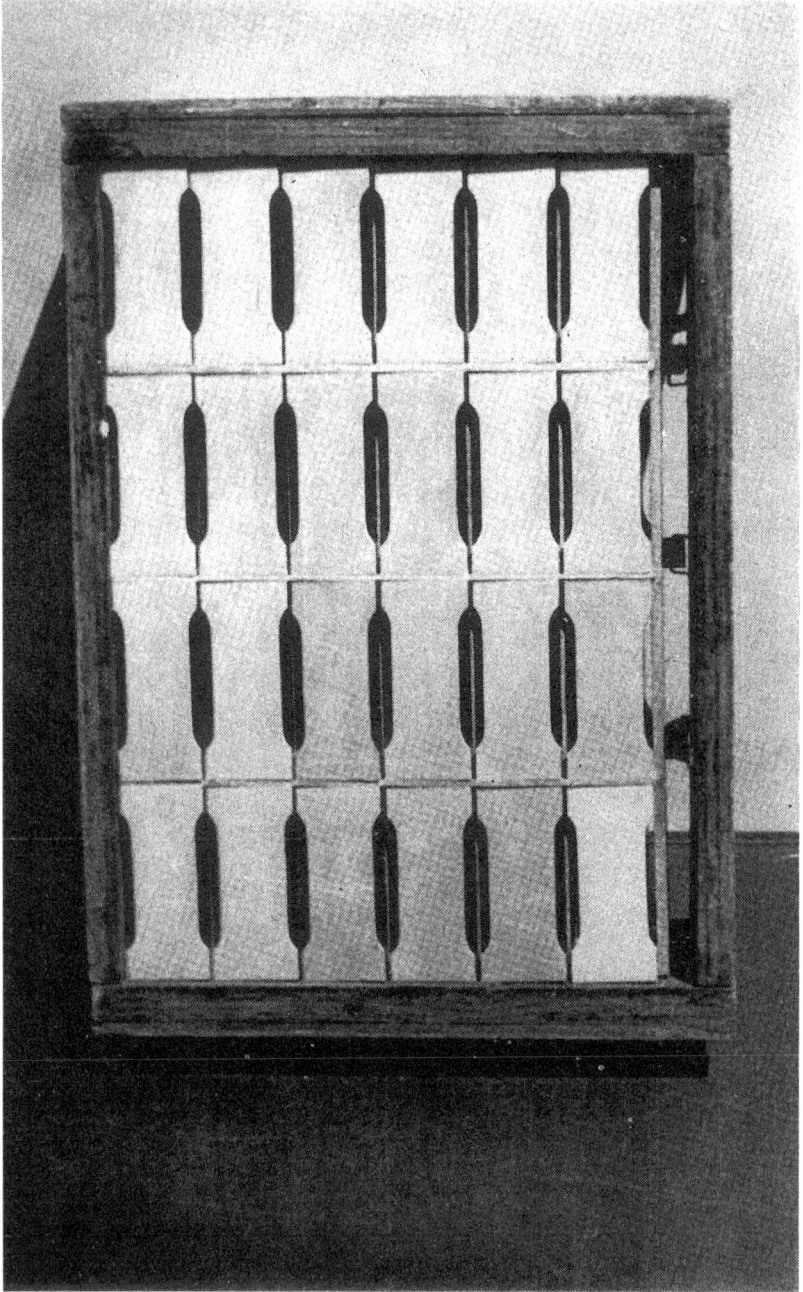

Fig. 12. The T super as used by Dr. Miller, with a follower board on one side.

We find less travel stain on the comb than in other supers, as the bee traffic is diverted to the sides and ends of the super. The elimination of travel stain to any degree is to be desired in comb honey. If the honeyflow is of average intensity, the corner sections will be completed as perfectly as the ones in the center of the super.

Fig. 13. The Miller super showing that end cleats were necessary to increase the length so it would fit.

All sections used in our ventilated T super are 4 ¼ x 4 ¼ x 1 7/8, two beeway and unsplit. Personally we do not like split sections and elsewhere in the book a description on foundation fastening will be found which will eliminate any desire for split sections. It might be desirable to get honey sections made from as white a basswood as possible as there is considerable less breakage.

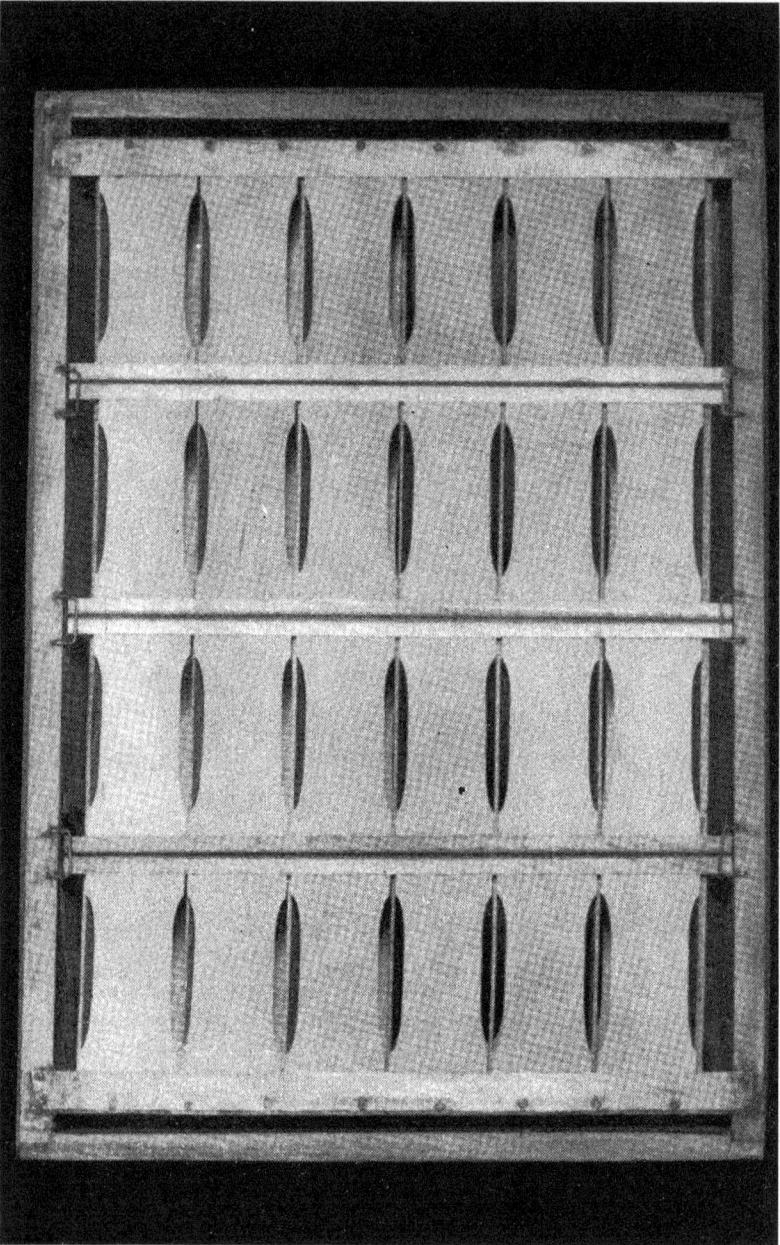

Fig. 14. Bottom view of ventilated super showing the T tins resting on the staples, and space for ventilation all around the sections.

Separators

The separators used in this super to separate the rows of sections are known as plain separators. These do not have any scallops along the edges like the separators used in supers having section holders. These separators rest directly upon the T tins and therefore scallops are not needed, as this arrangement provides a bee space at the bottom of the section. The width of the separators is approximately 3 ½ inches. This width will give the necessary bee space at the tops of the sections. There is a great saving in cleaning these separators at the end of the season as there is practically no breakage whatsoever. We have found considerable breakage in trying to scrape the propolis from the scalloped ones. If the small pieces between the scallops break off it changes the bee space and gives more places for bees to deposit propolis. The work is much faster in scraping the plain separators as a person does not have to worry about breaking off any little projections along the edges.

We tried making some separators of white poplar known locally as "hickory poplar." This is a fairly hard wood and is plentiful. The separators made from this wood handled very nicely and they appeared to hold up better than those from basswood. We notice the fiber of this wood does not loosen and gather on the scrapers as the basswood does.

We have bought regular scalloped separators and cut these to the proper size.

Sections for Comb Honey

Most sections are made from carefully selected basswood sanded smooth, 1/8 inch in thickness, grooved for folding, and dovetailed on the two ends for fitting together and holding when folded. Sections that do not fold squarely should not be used. It is important that sections be milled properly and that they be accurate in size, as those that are not full size will result in sections of comb honey being light in weight. One may find sections that are 1/16 inch too narrow; when this is multiplied by 24, or 28 if a ten-frame super is used, it amounts to as much as one section 1 ¾ inches thick.

Due to the scarcity of basswood it is not now possible to get sections made of as white wood as when this material was more plentiful. A slight discoloration of wood or not too dark streaks should not lower the grade of the finished comb. However, this does not mean to tolerate discoloration such as fingermarks, soot from smokers, and propolis stain.

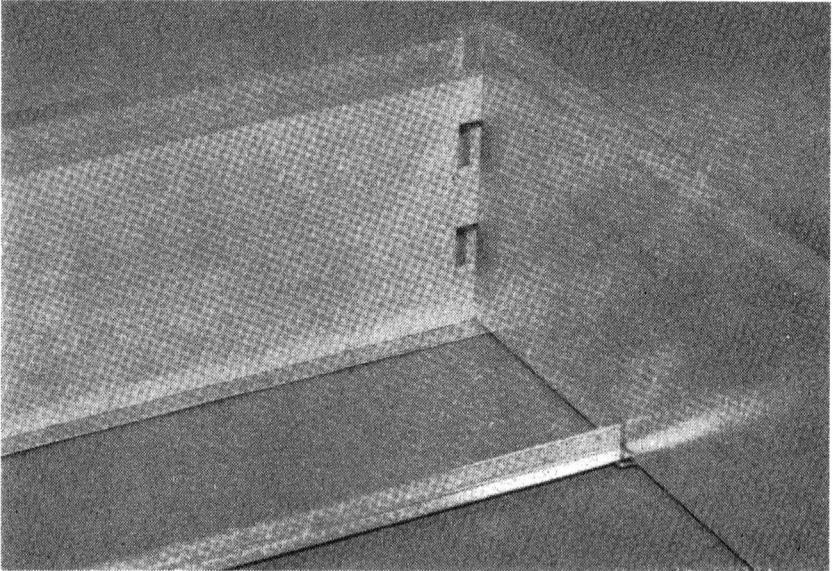

Fig. 15. Inside corner of the super showing openings for bee passage and for ventilation, and position of T tin resting on supporting staple.

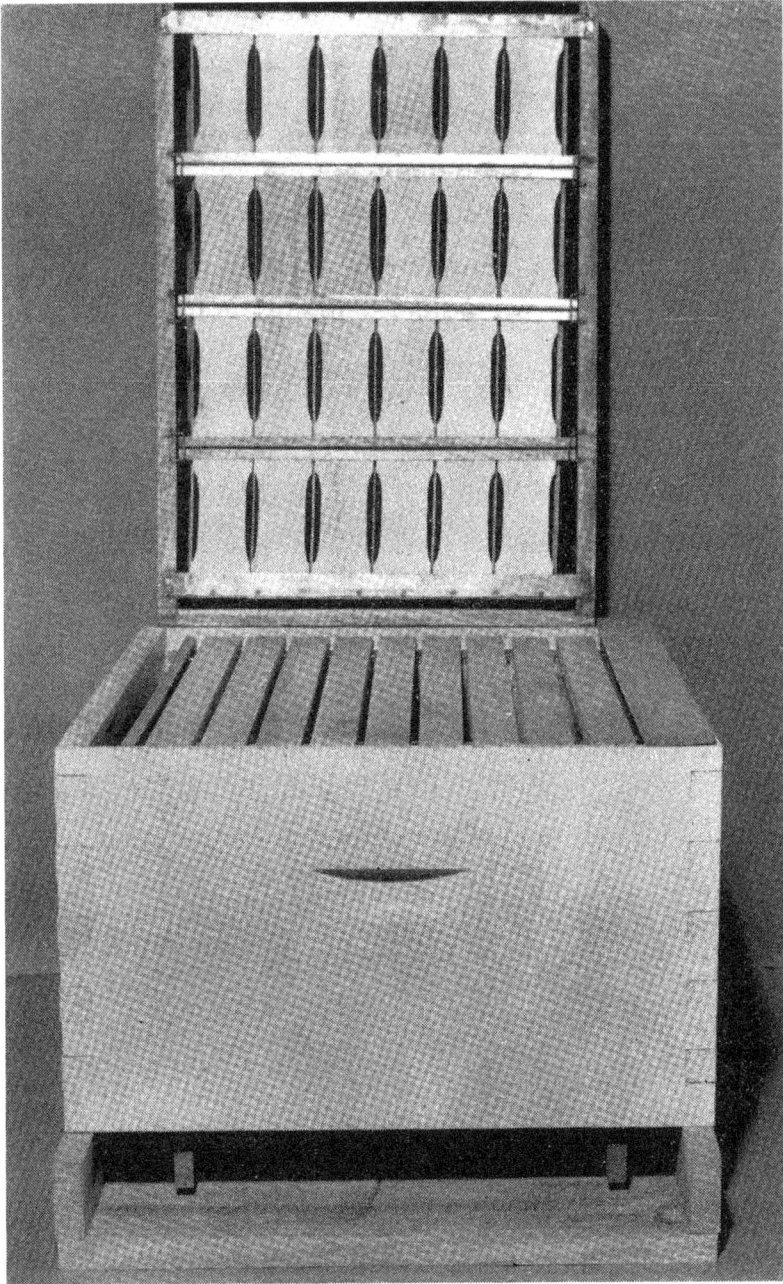

Fig. 16. The narrow super permits the sections to be directly above the brood at all times.

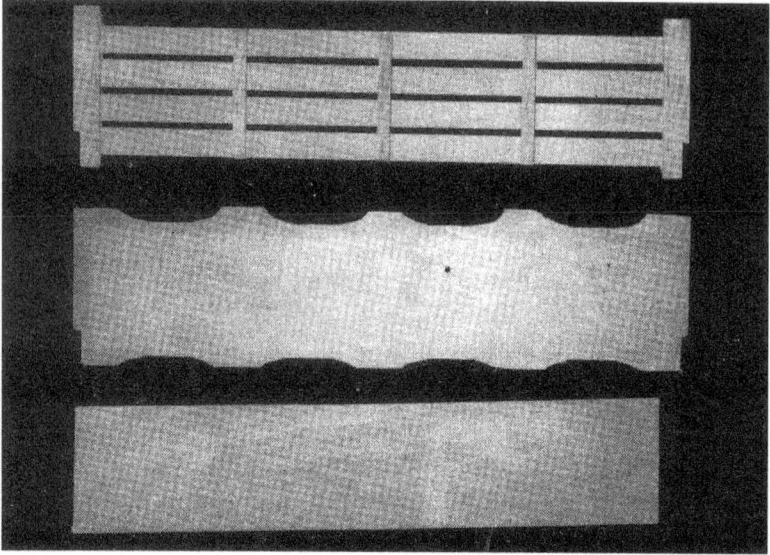

Fig. 17. Fence separator (at top) used with plain or no-beeway sections. Scalloped separator (center) used in standard supers having section holders. Plain separator (at bottom) as used in the Killion super.

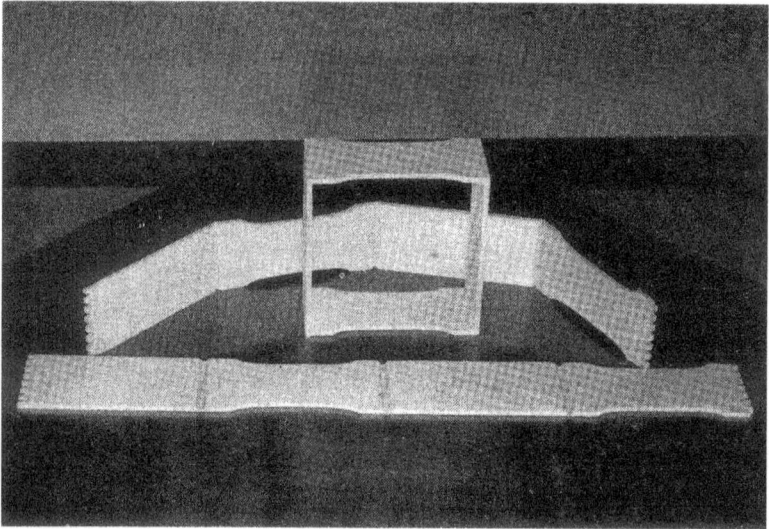

Fig. 18. Comb honey sections made from extra white basswood.

Fig. 19. A section press clamped to the table. The
author added the rack for holding the unfolded sections.

If some of our darker woods could be handled as easily as basswood, the white comb might show up more beautiful. Other woods might not fold as easily as the very white basswood.

The two-beeway section is scalloped along the sides of the top and bottom of the section allowing vertical passage of the bees. The beeway section having dimensions of 4 ¼ x 4 ¼ x 1 7/8 inches is the most popular and is used widely. This size section is the only one that can be used in the ventilated T super.

The 4x5 plain (no beeway) section has been used to some extent but not many are being used today. The idea of this size section was to enable the beekeeper to use it in the regular 5 3/8 inch extracting super. This would eliminate the use of two sizes of supers if both extracted and comb honey were to be produced.

The 4x5 sections require separators between the rows built of slats instead of plain or scalloped boards. They are called fence separators as they resemble a board fence. Upright pieces are fastened to the long strips where the edges of the sections meet each other in the rows. This provides a bee space and ventilation.

Folding Sections

Sections may be folded by hand or by means of a section press, by which the section is pressed together firmly and squarely. When sections are too dry to fold without breaking the boxes containing them should be placed on a damp concrete floor several days before the sections are to be folded. The floor may be sprinkled and the containers turned to allow the moisture to penetrate all of the V-grooves. A dampened rug or carpet thrown over the boxes is a great help, or the cartons may be sprinkled from time to time. When folding sections, there should not be any dry cracking sound and the sections should fold smoothly. I do not recommend pouring water along the V-grooves s the water will cause the outside surface of the section to become rough. The sections should not remain dampened any longer than required for folding as they may mildew, staining the sections.

The section press is a standard make. I added a rack for holding about 30 sections as I was always forced to remove some unfolded sections from my lap each time empty supers were needed. The rack is conveniently located just in front of the operator and sections are always within easy reach. A board is hinged to the top of the press which can be clamped to the workbench and makes the press more sturdy. The press can be easily removed and taken out of the way in a jiffy, if it is not needed.

During the folding of sections, one sits on a small stool in front of the press. An empty super, with an inner cover nailed on, is placed on the floor to the right of the press on which to stack the supers as they are being filled. The open carton of sections and the T tins are placed nearby on the table. The empty super is set on the inner cover and the three T tins placed in position. As the sections are folded they are put directly in the super. It is best to have all dovetails down when placing the sections in the super as it adds to the appearance of the section. We also turn the dovetails away from each end of the super, as later when the sections are being scraped, the scraper will not catch so easily on the dove-tails. About six supers can be filled before it is necessary to remove them to another place in the shop or until we are ready to fasten in the foundation. When the sections are folded the wood is somewhat damp and it is best to wait a couple of days before fastening the foundation as the drying of the wood may loosen the wax from the section.

Foundation for Sections

Only the best quality beeswax should ever be used for the manufacture of surplus foundation for comb honey. Beeswax will vary in toughness and the tougher wax should never be used. For many years we have been paying a small premium for our surplus foundation which has been made from carefully selected wax. If the beekeeper will use only the best foundation, he will not have complaints from customers about the toughness of the comb when eating comb honey. We often hear this complain made by people who have never eaten any of our comb honey. They would not mind buying comb honey, but they do not want a mouthful of wax to dispose of. We explain to these people about the careful selection of pure beeswax foundation used and guarantee them there will not be a "mouthful of wax left" If necessary at times, it is a good practice to cut into a section to show there is no toughness when the knife cuts through the midrib or center of the comb.

In ordering foundation for sections we ask that it be cut to a length of 15 3/8 inches instead of the standard length of 16 ½ inches. The latter length is suitable for shallow frames and split sections but for unsplit sections, 15 3/8 inch length is best. This size will allow the foundation to be cut into four equal lengths of about 3 ¾" inches.

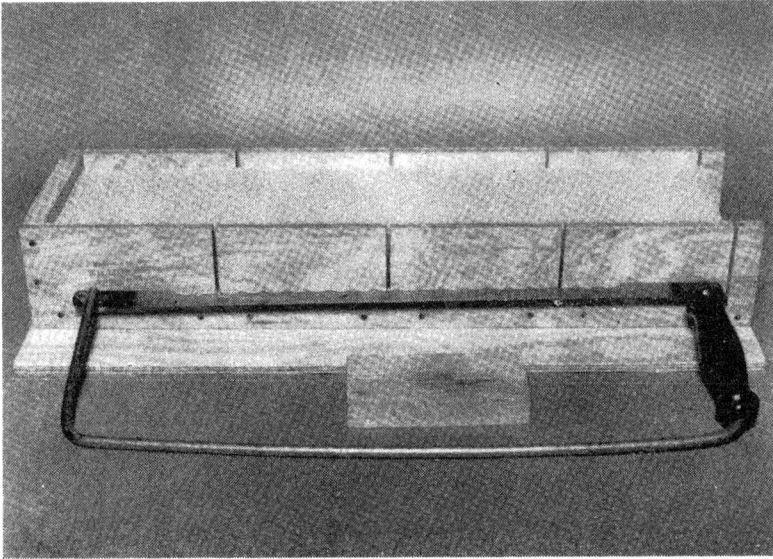

Fig. 20. Foundation cutting box with special cutting knife and block for gauge in cutting the small bottom starters.

Fig. 21. Foundation cutting box with block inserted in end for cutting the foundation into the two sizes.

Fig. 22. Small flat-wick kerosene stove with aluminum hot plate—the most economical and most suitable arrangement.

Cutting Box

A box for the purpose of cutting foundation the desired size may be purchased from a bee supply house or made in any beekeepers workshop. The box, for 3 7/8 width foundation, is about 2 inches deep, 4 1/16 inches wide, and the sides and bottom 16 ½ inches in length. The box can be nailed to a piece of plywood or Masonite somewhat larger than its base, in order to fasten it to the worktable to hold it in position while cutting. There are four saw kerfs in the sides but the one near the open end is not used except when the foundation is longer than 15 3/8 inches.

A thin knife, such as a scalloped slicing knife, is used with a sawing motion to cut the beeswax foundation. A better one can be made by getting a blade from a bread slicing machine and mounting the blade in a hack saw or similar frame. A temperature of near 70 degrees F. is about correct for cutting; if the temperature is too cool, the foundation will shatter along the cut edges.

The cutting box can be filled to within ¼ inch of the top. The foundation, held lightly by the left hand, is cut into squares. Then the foundation squares are removed and a small block, which is the width of the bottom starter, (5/8 x 2 x 3 7/8 inches), is inserted into the box against the closed end. The block causes the foundation to extend beyond the saw kerf the exact width of the bottom starter or 5/8 inch. As each stack of foundation is picked up to be placed in the cutting box again for cutting the bottom starters, it should be turned at right angles to its original position. This allows the foundation when placed in the section, to have the rows of cells running horizontally. It is very important to cut the foundation properly so as to have the rows of cells running horizon-

tally instead of vertically as many pictures of comb honey
will show. You will note the rows of cells run horizontally
in foundation for brood frames, shallow frames and split
sections. As the foundation is cut into large and small
pieces they are placed separately in cardboard boxes for
protection against breakage and to keep away from any
possible dust.

Sometimes it is necessary to cut a quantity of foundation in advance of the time it is to be used. When this
is done the boxes are sealed with tape and the contents
marked on the outside, "cut foundation, small pieces" or
"large pieces" as the case may be.

*Fig. 23. The foundation fastening board with one row of
sections in place.*

Fig. 24. Small starters in place ready to be fastened.

Fig. 25. Small starters fastened to the wood, except for one in the lower left hand corner.

Fig. 26. Large or top starters all fastened to the wood.

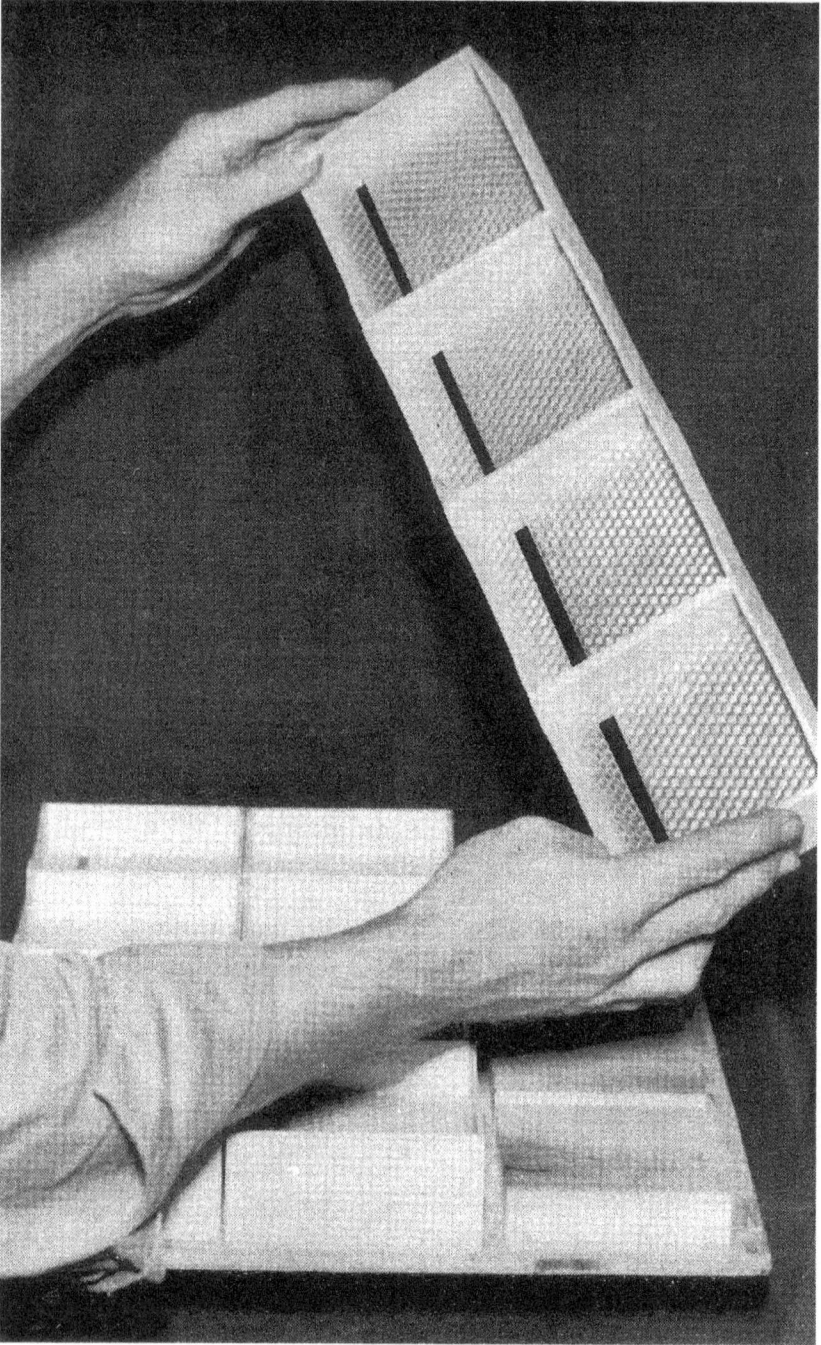

Fig. 27. The sections are lifted in groups of four.

Fastening Foundation in Sections

Most beekeepers dislike the job of fastening foundation in sections. The way they try to do it makes a difficult job. Some of them manage somehow to get the job done after a lot of fuss and complaining. The most satisfactory and efficient method for fastening the foundation in the sections is by the use of a multiple block board and aluminum hot plate. The twelve wooden blocks are nailed to a plywood board for a base. The blocks are three deep and four wide, 3 5/8 inches square, 7/8 inch in thickness and spaced 11/15 inch apart. The nailing is done through the plywood into the blocks, leaving the top surface of the blocks smooth. Before using, the blocks are painted with linseed oil. In warm weather a small amount of Vaseline smoothed on each block allows the foundation to slide across easily.

We find aluminum to be the best material for the hot plate. To use this hot plate, we have always used a small flat wick kerosene stove. An electric plate or even an electric heating device maybe preferred to our plate and stove. The stove is very economical to use as a pint of kerosene will last for many hours, and the heat is always delivered to the right place on the plate, without the bother of an electric cord which is sometimes in the way.

Our stove is equipped with two wires bent to form a U which prevents the plate from going too far down into the flue of the stove. The blade of the hot plate is 3 7/8 inches broad and 2 ½ inches deep. The blade should be heated only enough to melt the wax sufficiently for the foundation to adhere to the wood section properly. If the blade is too cool, the wax will not melt and stay liquid state long enough to stick to the wood; if too hot, the wax and the section may be discolored. It will require just a little practice to determine how hot to have the plate. If at any time the hot plate gets sooty or smoked, it must be wiped clean before using. If the room temperature is between 55 degrees and 65 degrees F. better and swifter fastening will result because the foundation starters are stiff and handle more easily.

The fastening board should be in position as shown in fig. 23 when placing the sections on it. The sections can be lifted from the super, or table, in units of four and placed on the board on the far side, and so on until the three rows are on the blocks. The sections, when placed upon the blocks, should have the tops farthest from the operator. The board should now be given one quarter turn counter clockwise; then the tops of the sections will be on the left side. The left side of the board can then be raised slightly and with a flipping motion the sections are forced to slide to the right, leaving the opening between the block and the bottom of the section. The small bottom starters are now placed on the blocks (Fig. 24) and fastened with the hot plate while the board is still in the same position. Holding the plate in the right hand, it is lowered into the opening between the block and the section. With the fingers of the left hand the foundation is pushed against the hot plate; as the foundation touches the plate, the plate is withdrawn quickly (not holding the foundation against it for more than an instant). As the plate is withdrawn, the foundation is fairly clicked against the section. Fig 25. The board is then given a one-half turn which brings it back to position as in fig. 26, except the tops of the sections are now to the right. The board is again raised on the left side and with a flipping motion the opening is made near the top of the section for the large top starter. After all the large ones are fastened, the board is again turned to position as in Fig. 27 and with a hand on each end of a row of sections (using a slight inward pressure), the row of four sections can be removed from the board, turning them to an upright position at the same time. As they are removed from the board they are placed back into the super on the T tins. The plate should be hot enough to fasten all twelve of the small or large ones without putting it back on the stove. Drops of melted wax which collect on the blocks and board must be removed or they will interfere with the sliding of the sections and starters.

Fig. 28. The center stick usually is inserted between the sections last.

It will require only a short time until a person is fastening foundation in sections at a speed that did not at first seem possible. A person can see the work at all times and any foundation not securely fastened can be seen at a glance and quickly corrected. A perfect job of fastening starters will eliminate crippled sections to a minimum. Foundation which is not securely fastened and falls out of bends out of shape will ruin the sections, as the comb will be out of shape or of drone cell size. Many times the writer has seen beekeepers using only a narrow top starter in order to save on foundation only to have the bees build drone comb instead of worker comb.

Filling the Super

To assemble the super with separators, sticks, and springs will require only two tools on the table, a hive tool and an awl or a broken ice pick. All separators should be the same thickness; 1/10 or 1/12 inch is best. The writer has both thick and thin separators and the thick ones are always used on the outside. After the separators are in place, one on each side and between each row of sections, the small sticks are placed between the rows of sections crosswise of the super. These small sticks are to fill up the space on the top side which was made by the T tin. This makes the sections perfectly square when in the super. We generally place the middle stick last. To insert these sticks between sections, they are started as in fig. 28. The two sections opposite the opening on the far side of the super should be grasped and pulled apart to make room for the starting of the stock in the opening, the next two sections must then be done as before at the same time using a finger to direct the stick into the opening. After all three sticks are in place, the super springs should be inserted, two on each side. The springs go in on the first side very easily but the hive tool must be used for the other side of the super. We space our springs an equal distance from each end and where the edges of the first and second section rest against the separator. Some separators bend very easily and if the spring were placed opposite the opening in the section the separator may bend and interfere with the comb in that section. Some times in putting in the springs one of the small sticks may be pushed down at one end; the pointed tool is for prying it back into position. A separator may b now used edge-wise to press lightly on the rows of sections to have a perfectly smooth surface. A smooth surface will prevent some propolis from being deposited on the tops of the sections. One may find considerable propolis if the sections are not perfectly smooth all the way across the super.

Painting Sections With Paraffin

Both the tops and bottoms of the sections in the ventilated super are exposed and should, therefore, be painted with paraffin. An electric hot plate has been found ideal for heating the paraffin as it gives the required heat and temperatures can be controlled. A five pound honey pail is an ideal container in which to heat the paraffin. A stiff piece of wire should be used across the top to wipe the brush on so the hot paraffin will fall directly into the pail and not run down the outside. A candy thermometer is used to control the temperature from 300 to 350 degrees F. Cooler than this will make the wax too thick on the sections and considerable more wax will have to be used. Hotter than 350 degrees will darken the wood and there is always the danger of fire. We use every precaution when painting our supers to prevent a fire. Additional fire extinguishers are placed within easy reach and our garden hose is attached to the water line in case it is needed. Care is used when removing or returning the brush to the pail and everything which may be combustible is moved away from the table.

Fig. 29. Painting the sections with hot paraffin.

A brush about 2 ½ inches wide is used for painting. A hook is made on the handle of the brush so it will not rest with the bristles on the bottom of the pail. The heat will quickly make the brush worthless if it rests on the bottom and the heat will make the bristles bend and not handle as well.

The super to be painted is placed on end upon the table and holding the top with the left hand it is inclined slightly backward. Before a super is painted with paraffin any dust particles should be brushed off with a broad table brush. The sections are painted a row at a time, down and up as long as enough hot paraffin is available in the brush for an even coverage. Most of the time the brush must be dipped in the pail for each three rows of sections. The super is then turned and the bottom painted in a similar manner. If there are a large number of supers to be painted with paraffin one should purchase the material in cakes of about 10 pounds each from oil companies.

Bait Sections

A bait super is a super having foundation in all the sections except one near the center which should contain empty comb. Each year a number of off-grade sections are robbed out by one apiary to provide us with sections filled with empty comb for this use. It is only the first super given a colony that has a bait section in it. Therefore, before the supers are assembled an estimate should be made of the number of colonies to be used for comb honey in order to determine the number of bait supers to prepare. The bait section is marked with an X scratched in the wood at the top so it can be identified without any difficulty and set aside. As the comb in this section will be darker and tougher than in the other sections in the super, we must never permit it to be packed with the other sections for market. These bait sections are all melted up at the end of the season for liquid honey. Some beekeepers try to use full supers of bait sections which is not a good practice. Queens seem to delight in going up into a super of drawn sections and laying eggs. We do not find this condition if only one bait section is used.

IV. Colony Management

IN producing honey, whether comb or extracted, the beekeeper should remember that his colonies should be strongest in field bees at the start of the expected honeyflow. This is important because each honeyflow usually lasts but a few weeks. George S. Demuth often said, "We must rear bees *for* the harvest and not *on* the harvest."

The beekeeper who produces both comb and extracted honey, should select for comb honey production only the strong colonies and those that may be expected to work best in foundation supers. The weaker colonies should be used either to strengthen other colonies or to produce extracted honey. Those who expect to produce comb honey entirely should make every effort to have all colonies in the very best conditions.

Spring Management

Some early spring management may be governed by the amount of honey left on the colony in the fall for winter stores. About seventy pounds of honey per colony seems to be the minimum here. All our colonies are in double brood chambers for winter and many will have nearer ninety pounds than seventy. A good supply of pollen should also be stored in brood combs. As our colonies are unpacked or if wrapped, as this covering is removed, each one is checked for stores and strength. Any dead ones are placed on the truck and brought to the shop, unless a dead colony has combs of honey which are needed elsewhere in the apiary. Every dead or weak colony is carefully examined for disease first even before being placed on the truck or used in the apiary. If colonies appear too weak to be of any value they are united with stronger colonies, as it does not help much to unite two very weak ones. Any hive needing food is marked and is given sugar syrup or a body with honey as soon as possible.

Good prolific queens in our hives in the fall should insure a good cluster of young bees which are so essential to good wintering and spring build-up. In wind-swept areas such as we have here in Illinois, the apiaries need some kind of windbreak, whether they are protected by packing cases or not. Our main fall flow is from hearts-ease which is excellent for winter stores. Most of the unsealed heartsease will contain less water content and yeast cells than sealed sweet clover honey. Winter packing, windbreaks, good stores, and a good cluster of bees all help to conserve energy which is so valuable to any colony in early spring. We do not put enough emphasis on that remark—conservation of energy. If our colonies have received careful attention in the fall and winter most of our spring work is just routine: unpacking, checking for dead or queenless colonies, some feeding, clipping queens, and giving an extra body of combs when needed during the dandelion flow. If weather conditions do not permit the bees to gather much from the dandelion, then we must feed sugar syrup to keep up brood rearing and maintain colony strength. Some apiaries require considerable pollen supplement.

Pollen Supplements

It would be unwise not to mention the value of pollen supplements. When Dr. Farrar first began making reports on the value of feeding supplements, I became interested. We were needing the very thing he had been working upon. Our apiaries, or most of them, were on the flat prairie where our first nectar and pollen source was dandelions. Some years our bees were able to store a surplus of dandelion nectar over what they needed for immediate use, and they gathered a good supply of natural pollen. We had some years when weather conditions were so unfavorable that bees were unable to work dandelion, except for a short period of time. If our bees failed to store up a surplus of pollen and nectar they were slow to gain strength for the clover flow. To our knowledge very little work had been done on pollen supplements or we would have tried feeding supplements sooner.

Our first experience in the use of pollen supplements proved their value, and we have continued to use them every year since. We have not found any complete substitute for natural pollen and I doubt if there ever will be one. We have, however, found suitable materials to mix with natural pollen. Our best results were obtained when we used at least one-fourth part natural pollen (by weight) with other ingredients. When we ran out of natural pollen, we used a mixture of five parts soy flour and one part each of brewers' yeast and dried skim milk. Some of our other mixes where made with dried whole milk. Later we used just soy flour and yeast when we did not have any pollen. Each year we fail to trap enough pollen. Hereafter we plan to use twice as many traps as in other years to insure us of enough pollen for our needs.

My advice to anyone interested in feeding pollen supplements is to trap their own pollen from disease free colonies. There has been some effort to market pollen but I do not approve of it, as I can see the danger of transmitting bee diseases by this method. We have used pollen traps and at the same time given pollen cakes inside the hive to the same colonies. These colonies appeared to continue gather pollen as much as those we did not give the pollen cake, and remained in normal strength.

Not all colonies will gather pollen alike. We find some will gather one certain color for several hours, another will gather one color for a while and suddenly change to some other flower as two colors will be present at near the same time. As a breeding experiment some-one should try to perfect a strain of bees which prefers to work red clover. In the pollen we have trapped, we have noticed some colonies have considerable more red clover pollen than others. It is only natural to believe some colonies prefer to work certain blossoms in preference to others. Checking the source of pollen which is caught in the trap may help in the selection of bees for pollination work. We do know that some colonies gather several times more pollen than others. It is a question at times why a certain colony will be working a particular kind of flower, when all the rest are busy on something else. Is it because the first pollen or nectar gatherers started on this flower or is it a matter of preference of this colony to work a certain flower? Do some bees have special flavor pref-erences that bees in another hive do not have? The way the bees act at times will lead one to ask many of these questions.

Selection of Breeding Stock

To those who follow the production of comb honey, and bee breeding for that purpose, are confronted with problems not to be found in regular breeding practices. The beekeeper who does not rear his own queens or a part of them, is missing one of the great thrills of bee-keeping. To me, queen rearing is fascinating, educational, and profitable. Any one of these three reasons would cause me to continue in this particular field.

In a review of all beekeeping literature available, the writer fails to find another person within the last twenty-five years who has made any effort of bee breeding for comb honey. Many breeders have made every effort to produce the best bee possible for extracted honey. In the selection of breeding stock or what we expect to have in comb honey stock, will carry us beyond the requirement necessary in breeding under present practice standards. Much of the carefully bred stock to be found on the market today is idea for extracted honey, but is not entirely satisfactory for section comb honey.

Our selection of breeding stock dates back almost thirty years. The stock we selected had already been used for this same purpose many years before that time. It is impossible to keep the same bloodlines over many years, but selective breeding is a never-ending job. Good and bad stock is always present due to uncontrolled mating. It is always our job to retain the good qualities and to eliminate the undesirable ones as quickly as can be done. In our list of requirements we expect all virtues required by the man who produces extracted honey, plus a few more. It does not mean much to the man who slices off the cappings whether they are uniformly white, of even pattern and design, if the cappings are raised from the surface of the honey or lay flat, or if they are evenly attached and sealed at the edges of the wooden part of the frame. These are only some of the things to consider in breeding for comb honey. We want bees that will attach the comb evenly to the top, bottom, and sides of the section; bees that will build full thickness comb in the sections and not brace them to the sides of the separators; and bees that do not propolize the sections too extensively. Bees that accept the entire super and do all their work uniformly are the ones desired; not those which start only on one side and work across to the other side.

Later, the results on the chart in the shop in comparison with reports shown in the record book, plus our own personal observation, are used in the selection of our breeding queens.

One age-old remark that we often hear, which should be exploded, is "Black bees are superior for comb honey over Italians." The claim is made that they cap the honey whiter. In the many years that I have been engaged in beekeeping and apiary inspection, I have examined many thousands of colonies of bees. I will make an emphatic statement that during all that time I have never seen a colony of black bees which would equal any of our best Italians in comb finishing. We do hear some good claims for the Caucasian bee for comb honey. My experience with two strains of Caucasians was not favorable.

Our season for rearing queens is only during the swarming season which is during our honeyflow and the weather is warm. In view of these conditions we have adopted a baby-size queen rearing nucleus hive. If we were to practice queen rearing over a longer period we would want a larger size hive to eliminate any possible chance of queens being chilled before merging. We have used these small hives for five years and we like them. The original cost is not so great as those which would take regular brood frames. The upkeep is rather low and the small size does not require as many bees to fill them. About a quart to each nucleus makes about the right amount for our conditions. The frames for this hive are made of four 4x5 honey sections ripped to a width of 1

1/8 inches. These four frames are spaced with the small top bars resting in notches along the tops and sides. The feeder is placed beyond the frames farthest from the entrance.

Our queen rearing yard where these small hives are used is isolated very well as there are no other bees within a two and one-half mile radius excepting our own. To this queen rearing apiary we move two to five colonies which have been carefully selected for the rearing of drones. These colonies are supplied with drone-size foundation to insure an adequate supply of drones for mating purposes.

Since we have established this isolated queen rearing yard, we have noticed a decided change in our breeding results. Our wall chart gives about 200 per cent gain in exhibition grade of honey.

Swarm Control and Requeening

Perhaps the problem of swarming has done more to prevent beekeepers from producing comb honey than all other reasons put together. Swarming, if not controlled, does ruin the chances of getting a good crop. Some years bees are less inclined to swarm than in other years. We have had them almost refuse to swarm some years, while in other years, with almost identical conditions, they swarm galore. At the start of the swarming season we do not know how the season of swarming is going to be. Whether the bees themselves know—very far in advance—what kind of a swarming season it will be, I do not know. It would be interesting to know whether they do have such advance knowledge.

Fig. 30. All colonies are wintered in double hive bodies.
Weak colonies are united with stronger ones.

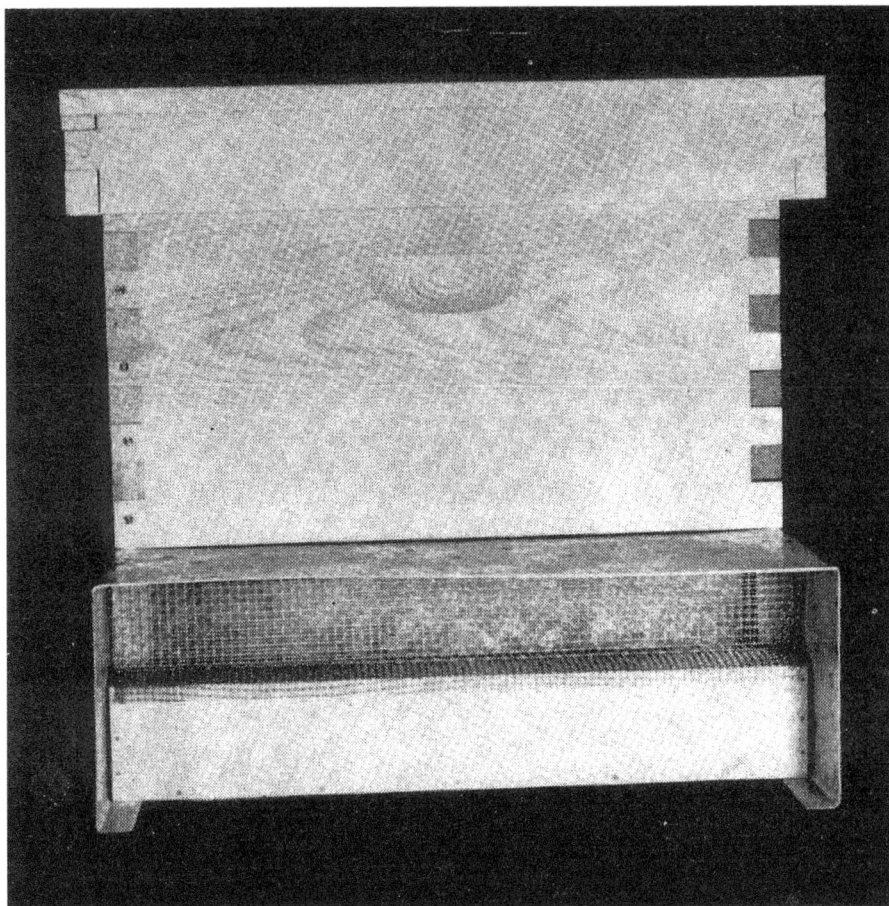

Fig. 31. Pollen trap on hive.

There are many methods of swarm control. We have tried many of them, each with varying degree of success. There are some controls used in extracted honey production that cannot be applied to comb honey production. One will see some clever ideas in print from time to time on ways to prevent swarming. Some of these are excellent for a few colonies but unsuitable in commercial production. There are extra gadgets to let bees fly out one side for a few days and out another side over the weekend, but if we plan for extensive operation, such things as screens, queen excluders, hive bodies and numerous other articles must be eliminated.

Fig. 32. The result of good breeding stock.

We have a system that we have used now for over 25 years and every year we like it better than ever. The practice of clipping queens is something we must do. There is no other alternative or we will lose our bees.

In our method of swarm control we are trying to do three important things at one time: control swarming, requeen our colonies, and get a honey crop. In doing all these at one time calls for precision on timing our work to correspond with all the details of the three projects. All three are so correlated that it is impossible to work them as separate projects.

Fig. 33. A queenright colony with space prepared to receive a bar of queen cells.

Fig. 34. Giving a cell bar of thirteen grafted queen cells to the queen right colony.

Early in my beekeeping career I learned to rear queens. To me, queen rearing is one of the most fascinating and educational phases of beekeeping. Any beekeeper who has not had the pleasure of grafting larvae into queen cells and day after day watching the development until finally the queen emerges from the cell, has missed one of life's rare treats. I have the fullest respect and admiration for our queen breeders; knowing they are performing a real service to the industry. What one must realize and understand is that the queen breeder must rear queens from early spring until late fall. He cannot select a certain choice period and then quit for the rest of the year.

Beekeepers who wish to raise their own queens can select the most suitable time in the season and have the finest queens possible. We do not practice queen rearing or requeening throughout the entire season. We try to rear our queens under the most ideal natural conditions to be found—*under the swarming impulse, in queenright colonies, but from larvae grafted from our finest breeding stock.* No other method in the world will give any finer queens.

I do not agree with the idea of making a colony queenless and then giving it a bar of grafted cells. The poor bees are then trying to build these cells under a strain of necessity. Our experience in giving a queenless colony a bar of cells is that they hurried the job along too much. We do not believe they give the cells as much jelly as bees do in a queenright colony. At least we know the bees will seal the cells quicker (according to our records) in queenless colonies than in queenright ones. The bees appear to sense their plight and want to remedy the situation by trying to get a queen as soon as possible. I started to say the bees "think" and we are told that bees do not think. Much has been found out recently about the way bees communicate with one another, some day we may be able to fully understand their language.

We kill all the queen cells we find on our first visit which is usually about four days after cutting our colonies down to give the comb super. Our reason for destroying all the cells is that the swarming impulse is not as intense as we want for good cell building.

Two, or maybe three, days later we may return to this apiary and find the colonies building queen cells with intense eagerness to swarm. We now start our swarm control method and our queen rearing program.

Again we kill all queen cells in a colony, remove the two follower boards, and spread the frames apart near the center of the brood nest. This space is for the cell-bar frame we expect to give this colony a few moments later. We save the jelly from the queen cells that were destroyed. As we are going through this colony killing cells, we also make sure the *queen is present.* The super or supers are returned to the hive and the cover placed on top, as we prepare our bar of queen cells.

These bars hold 12 to 15 cell cups, depending upon how they were placed on the bar. In most cases we now have 15 on each bar and we practice staggering the row instead of placing the cups all in a straight line. We have used the wood cup to hold the wax cell, it does not give us any better queens but we like to handle our cells on these wooden cups.

If two of us are working in the same apiary, one will prepare the bar of cells with royal jelly while the other is getting the frame of brood containing larvae of the proper age to be used in the cells. We like to use the fresh jelly just collected from the last hive we were working, the one in which we prepared to receive the bar of cells. The cells must be perfectly clean and not contain any dirt of dust. The tool we use for placing the jelly in the cups is made from a steel crochet hook. The hook part being cut away and the end flattened slightly. We dip some diluted jelly into each of the cells, using an amount about the size of an ordinary wooden match head. The jelly may be diluted slightly, either before it is placed in the cells or afterwards. We want the jelly to appear like thick coffee cream when placed into the cells or flatten out somewhat in the bottom of the cells. If the jelly is too thick when placed in the cells it continues to dry out rapidly and in a short time does not serve as a cushion of food when the larva is placed on it.

The frame of brood taken from the colony where we have our breeding queen, is to furnish the larvae. The larvae selected are as small as possible or less than twenty-four hours old and as uniform in size as is possible to obtain. They should be fairly floating in food when each one is removed from the worker cells with the grafting tool and placed on a cushion of royal jelly in the cell. We try to place the larva in the center of the cell in a natural position. If the larvae to be grafted are not surrounded by plenty of food they cannot be lifted from the cell as readily as those with plenty of food. If they appear to be short of food it is best to select the larvae from another colony. In case there is any doubt about the age of the larvae to be grafted one may select them about as big around as the smallest lead from an eversharp pencil.

Some queen breeders use a grafting needle with a hook on it for hooking into the curled part of the larva. Others like a needle with a small flat paddle. I personally like the latter. When getting the larva on the end of the needle we try not to put the paddle too far under the larva, but just far enough so when lifted out of the cell, one-half or one-third of it is hanging over the side. When placed on the jelly the larva will adhere to the jelly and the grafting needle is gently removed

As soon as the last larva is placed in the cups, the bar is fitted into a special frame and placed in the hive we have prepared.

We have removed this colony's own swarm cells and are giving it a single bar of fifteen cells from our best breeding stock. When the super is removed we will see the open space we left now filled with young nurse bees eager to start feeding these cells. This is the most ideal condition to be found for building cells; they are built under the swarming impulse, but from our selected breeder.

The bees will focus their attention on those cells and they will produce queens of the highest quality. These are queens that can be depended upon to last longer than one season. We graft several bars of cells in this apiary on that date. For example, we will suppose this was done on June 1; on June 3 we come back to this apiary and start killing queens. All queens are destroyed except those in our cell-building colonies. About June 7 we will destroy any queen cells the colonies in this apiary have started in the ones that are queenless. On June 10 or early on June 11 we remove the bars of cells we grafted on June 1. We give one of these cells to each of the colonies made queenless on June 3, but not until we have destroyed every queen cell they have built.

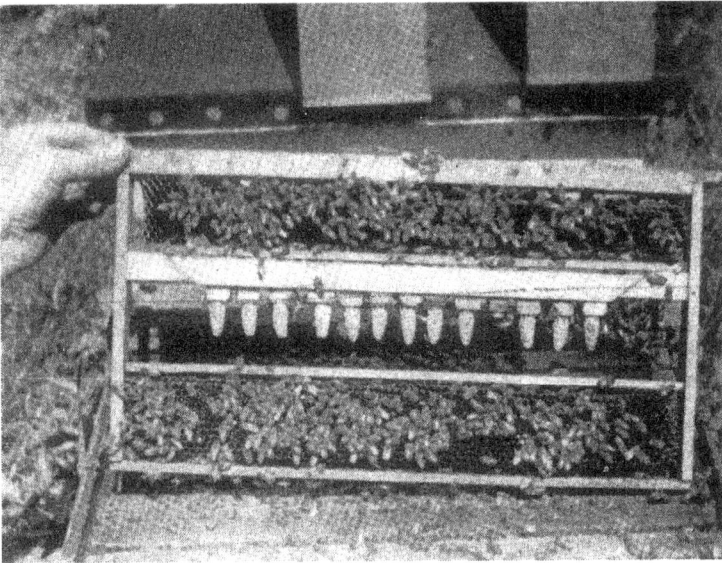

Fig. 35. A bar of good queen cells.

Fig. 36. A small mating hive.

We dare not leave a single queen cell except our grafted one. If we do, this colony will surely swarm. When they do the fine virgin will be lost along with the bees. This loss will also mean our honey crop from that colony is lost.

To find all the cells on the combs we must shake the bees from the combs in front of the hive. The combs will be full of fresh nectar and care must be used to not shake too much of it out of the comb. The bees seem to enjoy making queen cells in unusual places. These last cells will be small and may look somewhat like a drone cell. If we only have the grafted cell left in the colony we can rest assured that that one will not swarm. We have had only two cases of colonies swarming when only one cell was left, another virgin may have entered the hives in question.

The queen that was left in the cell building colony will swarm too about the time the cells are sealed. At this date we are ready to kill this queen anyway, and afterwards this hive is handled as the other queenless hives. It will be given a queen cell after being queenless eight days.

About five days after we give a bar of cells to a hive we examine this hive again to kill any queen cells they may have built on the combs. This is to insure us that we did not overlook a cell on the day we grafted the bar, otherwise, it may emerge and the colony will swarm. We must be careful not to shake the cell-bar frame when handling these colonies.

For many years we gave just queen cells to requeen all our apiaries. For the past five years we have been using small queen mating hives to save surplus queen cells. We then give these young mated queens instead of the cell. We like to have this surplus of young queens available.

During the queenless period the bees do not do too much in the supers, the brood chambers however become filled with new honey. When the young queen mates, this honey is moved rapidly to the supers, to make egg laying room for the young queen. It is from now on that we get our finest comb honey. It is impossible to crowd them enough to make them swarm.

Dipping Queen Cells

Almost every commercial queen breeder has his own pet method of dipping cells. Our method is not so elaborate as theirs, as we do not need the large number that some breeders use. Our cell forming sticks were made by one of my brothers from a small tree called ironwood growing in Indiana. I have never looked up the botanical name; just content to use the common name so familiar to all Hoosiers. The wood makes excellent cell forming sticks as the cups may be removed easily. The wood does not swell and fuzz up like some wood. About one half inch from the end, the stick is slightly larger than 5/16 of an inch in diameter and tapered slightly to the end which is rounded off on the edges. About the time the wax is starting to heat, five of the sticks are dropped into a glass of water. One is later removed from the glass, the excess water shaken off and dipped into the melted wax to a depth of about 3/8 inch. The stock is pulled out and turned back and forth a few times to roll the excess hot wax around the base to cool. The stick is again lowered into the wax but not as far as the first time. It is again twirled until the wax has cooled enough to lay the stick down and start with a new one. It will take four dips with each stick to give the cell a proper heavy base. Each dip into the hot wax must be shorter. In this way the open end of the cell cup will be very thin, but the base will be thick and protective. After all the stocks have received the four dips, the cells may be removed by a slight twist as they are pulled from the end of the stick.

Many large queen breeders mount several sticks in a row on a narrow strip and dip all into the wax at one time, shortening the length of the dip each time to get the same results.

We use wooden cups into which we stick the wax cell for ease in handling. These wooden cups do not help in the quality of queens. Their use is a little extra work, but we like them.

Clipping Queens.

My first experience in trying to clip a queen's wing was most amusing. I wonder sometimes if other begin-ners were as nervous as I. After a least a dozen at-tempts, I succeeded in picking up the queen by one wing. She buzzed around in a circle, and being afraid the wing would be twisted off, I released her on the frame. Again I started my pursuit of her in the same manner as person trying to pick up a mouse by the tail. This was a colony of black bees that had been transferred to new frames. There were several holes through the comb and this queen delighted in playing hide-and-seek from one side of the comb to the other. The instructions had been careful-ly read and the pictures did not show anything to compare with my problem. Finally she was again picked up and carefully placed between the first two fingers of the left hand and the thumb pressing lightly against or on top of her thorax. The queen continued to squirm around like an angle worm, but the clipping was done. My! What a sigh of relief when it was over.

The next queen was not pure Italian but large and yellow. She was picked up more easily and in placing her in position for clipping, she grasped my left forefinger with all six of her legs. My thumb went down gently on the three nearest legs and presto, she remained so calm that I proceeded to clip her in that position. The job was done so easily that I have never changed my method of clipping. All the pictures and descriptions of how to clip a queen are different from mine and I wonder why my method has not been described before. There may be hundreds of others who held a queen exactly as I do but so far, I have not seen any mention of it.

The Honeyflow

There is not much difference in the spring management of colonies intended for comb honey than that used for production of extracted honey until the flow starts. It is the desire of every beekeeper to bring his bees up to the peak of strength near the start of the main honeyflow. This desire cannot always be realized. All our colonies are in double brood chambers, sometimes they may occupy three if dandelion yields good or we feed pollen supplement rather heavy. Regardless of whether colonies are in two or three hive bodies at the start of the honey flow, *each colony is always reduced to a single hive body when the first comb honey super is given.* Up to the start of the honeyflow we practice reversing the two brood bodies about every seven days to help equalize the brood rearing conditions in each body. After the flow starts and fresh nectar is plentiful in the bodies (when the combs are heavy and the nectar drips easily), we place one hive body of brood, bees and queen on the bottom board that we intend to leave for the comb honey and give one comb honey super containing the bait section. In most cases, it is the upper body if two were used, and the middle one if three bodies were used for the building-up period. We do not sort combs to select any certain age brood, but try to have as much brood as possible in the hive body with the queen. Most of the bees are shaken from the hive body taken away, leaving only enough bees to take care of the unsealed brood. The bees are shaken from the combs right at the entrance of the hive body containing the queen. This extra brood and few bees can be used to help some weaker colony, which may later be used for comb honey or used to fill these extra bodies with honey for winter feed. The body of brood may also be used to make increase; if such be the case it is given a hive number and may be built up with other bodies of brood until it is two, three, or four hive bodies high. All our increase is made at this time of the year, limiting the number of new colonies by how many hive bodies we use in each colony of increase. Each new colony is well supplied with brood in all stages, enough bees to feed larvae and plenty of food. Two days later this colony of increase may be given a ripe queen cell or a young queen. It is surprising what enormous colonies these will develop into. Sometimes these stacks of increase are later "cut down" for comb honey and they are excellent for this purpose.

The colonies prepared for comb honey and given the super will, in most years start building queen cells in a few days. It appears that we have done everything to encourage swarming. We have reduced a large powerful two or three story hive to a single story and given them most of the bees; the single brood chamber is fairly jammed with fresh nectar, and we give them a super all cut up with small compartments (the sections).   It isn't any wonder they want to swarm.

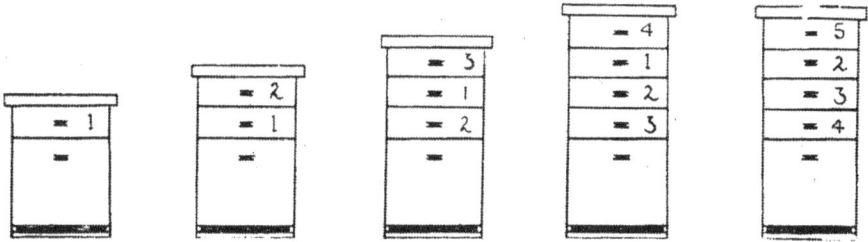

Fig. 37. Diagram showing the method of supering. The supers are numbered in the order in which they are given.

Super manipulation

When the first super is given, the bees should start drawing the foundation at once, depending on the intensity of the flow and colony strength. The first super should be one-half or two-thirds full before the second super is added. When a new super of foundation is given, it should almost always be placed on top of the other supers already on the hives. On the next trip the second super is placed next to the brood, and the first super is placed on top. If the colony is ready for a third super it is placed on top of the other two supers. Before giving the third super, the first super should be nearly full and the second one at least half full. In an exceptionally heavy flow, the third super can be given before the second one is one-half fill, but it is better to crowd the colony just a little than to give it super room too far in advance. One serious mistake made by many beekeepers is to give comb honey supers too fast or too many at one time.

If the next trip reveals that the third super is being drawn rapidly, it is placed next to the brood with the second super above it, and the first super on top. If the fourth super is needed, it is added on top. As soon as a super is completely capped over or finished, it should be removed. This is to eliminated all unnecessary travel stain on the cappings, and to avoid the handling of this super with each colony manipulation. The next visit may necessitate the giving of the fifth super and the removal of the first super. The proper order of supering is shown in Fig. 37. It should be advised that comb honey supers cannot be staggered, leaving openings for additional ventilation as practiced with extracted honey supers. Such practice will cause the bees to leave sections unfinished near these cracks or openings.

For some reason known only to the bees themselves, they prefer to work a little better in the back two-thirds of the super. The front one-third is almost always slightly behind the rest of the super. In the manipulation of supers we are constantly reversing the supers that are nearing completion, putting the front or lesser filled at the back. By this reversing of the ends of the supers, the sections are uniform in appearance when completed.

The examination of the super from the top side does not always show when it is ready to be removed. The super must be viewed on the bottom to make certain the comb in the sections is completely sealed. One may think the super is ready to come off the hive by just looking down on the comb from the top. The top view may show the section to be completed and yet several rows of cells near the bottom may be unfinished. When a super is found completed and ready to be removed it is returned to the top of all other supers. We now give several puffs of smoke into this super to start most of the bees downward. The super is grasped firmly and many of the bees dislodged by a few shakes in front of the hive. If more than one super is to be removed each one is handled in this manner. The escape board is now placed in potion on the supers already on the hive and finished super placed on it and then the cover if bees are in a partial shade. If hives are in the sun, a ventilator board, or ventilated rim is put on the super before the cover, making sure not to leave any openings above the escape board. Any openings where bees can get through must be closed, for bees are quick to rob out supers of honey above an escape if the flow would suddenly slow up. We prefer to use the Porter bee escape to any other method of getting the bees out of the supers. We find very few bees left in them 24 hours after placing the supers over the escapes. During the extremely hot weather it is necessary to use a ventilated rim between the cover and the supers above the escape board to prevent the possible damage of the comb melting and smothering the bees.

As each super is removed from a hive, the hive number is scratched planning on the top of one of the sections. In the scraping and grading of the crop, we are able to have a complete production record of each colony. A wall chart having all the hive numbers with a space for checking production is placed conveniently near the grading table. Each colony is credited with all marketable sections and a mark for certain outstanding qualities. The chart is valuable in checking colonies for desirable characteristics for use in our bee breeding.

Preparation of Bees for Winter

Very few beekeepers in this latitude make much preparation for winter protection of their colonies. We believe in making the necessary protection for our colonies to conserve as much food and energy of the bees as possible. In providing this protection, whether it be windbreaks, wrapping with tar paper or the use of a winter packing case, we can expect our bees to survive winter better and build up better in the spring.

Three major requirements for good wintering are a good cluster of young bees, an adequate supply of good quality honey, and sufficient pollen. When I say good quality honey, I mean honey well ripened or honey of low moisture content. We prefer equal amounts in weight of unsealed heartsease honey and sealed aster honey, as we find less moisture in the unsealed heartsease than in any sample of the aster honey so far examined.

To have a good cluster of young bees in our colonies in the fall means that we must have a prolific queen. If we have an old worn out or failing queen, we cannot expect a large number of young bees in late fall, and the same holds true if a colony was requeened too late to give the young queen a chance to lay a sufficient number of eggs to produce the necessary amount of workers. Any colony with a large number of old bees does not winter well. Much of our success in wintering is determined by weather and floral conditions during the latter part of August and the month of September. It is during this period we get our major fall flow and can expect our peak in brood rearing.

Colonies run for comb honey will under normal conditions have the brood chambers much heavier with honey in the fall than colonies which produce extracted honey. Sometimes there is danger of having too much honey in the brood chamber during the early part of the fall flow as this will restrict brood rearing. Years ago the writer would continue to take most of the heartsease crop in section comb honey as there was a ready sale for this type of honey in some areas. In recent years instead of using our comb supers, we tried putting on shallow chunk supers or full depth hive bodies. It gives us more room for brood rearing as our brood chambers are less congested with fresh nectar when using these frames than when using comb honey sections. For the past two or three years our winters have been very mild. It is true our bees did not require all the extra care given them for winter protection, but ever so often we get one of those winters which people call "old fashioned" when we have sub-zero weather for many days, with deep snows and howling winds. When one of these winters shows up, we see the unprotected apiaries suffer a heavy loss. I do not know the average winter loss for Illinois for the winter of 1935-1936, but it was well over 30 percent. During the winter of 1939-1940, Illinois had over 35 per cent winter loss with many larger apiaries showing as high as 75 to 95 per cent loss. Any state with such losses has a wintering problem.

We have for many years followed the practice of packing our colonies for winter. Bees can be overpacked as was the case when the old quadruple colony packing case was recommended several years ago. As we have been steadily increasing our apiaries over a period of years we find it impossible to continue building our winter cases and find it necessary to substitute the packing by wrapping some of them in tar paper. We perhaps will never find anything to replace the case we have been using.

The collapsible cases the writer uses are patterned after the ones used by Chas. A. Kruse. The case will hold two double-story hives either eight- or ten-frame size. The case is quickly assembled or dismantled. Regular ¾-inch lumber is resawn to a thickness of about 5/16 inch and this is nailed to a frame support and covered with tar paper. We have used tempered presswood without the tar paper and this material works excellent except it will warp unless supported well with a framework.

Fig. 38. Hives with winter entrances in place.

Fig. 39. Hives removed from the hive stands and the
bottom of the packing case in place.

Fig.40. The hives are placed on the bottom of the packing case and moved close together.

Fig. 41. Parts of the winter packing case before being nailed together.

Fig. 42. The assembled winter packing case ready to be placed over the hives.

Fig. 43. The winter case is now ready for the packing material.

During the summer our cases are stacked in a corner of the apiary. We assemble them near this place and one man can step over inside the case and carry it wherever it is to be used. As one person is distributing the assembled cases another can be nailing them together. Only four seven penny box nails are needed to hold the case together, one in each corner (inside).

One undesirable thing we find in using this case is that all hives must be handled twice in packing and unpacking. It does, however, enable one to determine the exact amount of winter stores each colony may have.

Fig. 44. The cover is wired down at each corner to prevent it being blown off.

For over twenty-five years our hive stands were made of 1x4 inch material. These stands were 20 ½ inches wide and 38 inches long, reinforced in the corner with a block of wood. These stands will hold two colonies which must be set off the stand and our winter case bottom placed in position. The colonies are placed on this case bottom and the packing case lowered over them. Two men can do this much better than one. For packing material, we find baled wheat straw is the most economical to use and the most plentiful. One bale will pack an average of three of our cases or six colonies, sometimes a little more, depending upon the size of the bales. Two of our apiaries located in a wooded area are packed with forest leaves. After packing material is placed around the hives, a cover is placed on the case and wired down at each corner.

Over a 20 year period our average winter losses in these cases have been low, from 2 to 4 percent. One year we had a 10 percent loss. Lately we have tried a new hive stand and for our purpose it worked excellently. We have a manufacturing plant in town that has a surplus of hardwood rippings. We use two strips 38 inches long and three short pieces 23 ¾ inches long. Holes are bored near the ends of the two long pieces and two short ones. A large galvanized nail is put through these holes and bent on the underside with the other short crosspiece placed in the center for support and fasted exactly as the rest. This hive stand can be folded almost in a straight form and hauled easily. When ready for use they are simply unfolded and two hives placed upon them. This hive stand will also permit us to use the hive lifter when moving, and when colonies are wrapped in tar paper, we can pass our twine underneath the hives to hold the paper in place.

Fig. 45. *They do so much and receive so little. Surely they deserve more than just the food they gather.*

Cleaning of Bee Equipment

No matter what style of super is used to get a crop of comb honey, they should all be cleaned before placing new sections in them for the next harvest.

We find fall and winter the best time to clean our supers and fixtures of propolis as the propolis scrapes better when the temperatures is cool. The top and bottom edges of the supers, as well as the inside are scraped clean each season after they are used. For scraping the separators, a board about six inches wide and two and a half feet long is placed on the table, one end extending an inch or so over the edge to facilitate the work. The flat scrapers used for this purpose may be purchased at any hardware store or made in the home workshop from old handsaw blades. The small wedge-like sticks used between the tops of the sections are scraped on this board.

For cleaning bee escapes, T tins, and super springs—these are boiled in hot lye water or similar detergent. We are now using sal soda and like it much better than lye. For boiling the T tins, we use a piece of one-inch mesh poultry netting about 36 to 50 inches long with a wooden strip nailed to each end to grasp. The netting may be twelve or fifteen inches wide. With the netting arrangement each end can be raised or lowered while the T tins are in the boiling solution. This movement permits the water to reach all parts freely. After boiling, the netting with the tins is lowered into a tank of clear running water to rinse and then placed on end to dry. The T tins must be dried quickly to prevent rusting. The writer always prefers to do this work in a well-ventilated room or out-of-doors on the concrete driveway.

Fig. 46. Wire netting to hold the T tins when boiling in lye water.

An ordinary wash boiler may be used for the tank to boil the metal parts, but it requires a good hot fire to keep the water boiling as the T tins or whatever is being cleaned cools the water each time a fresh basket full is boiled. To a boiler filled one-half full of water, it will take about three cans of lye or similar material. As the work progresses, more of this material must be added to the water. If the T tins are covered with a thin yellow film of gum, it is a sign the water may be too cool or the detergent is too weak. The tins and other metal should come out of the water bright and new looking.

Fig. 47. Hardware cloth baskets for boiling super springs and bee escapes in lye water.

Each winter we can find plenty of shop work to keep us busy. The scraping of propolis from the supers, separators, small sticks, escape boards and the boiling of metal parts must be done each year.

Our small mating hives must be checked and new frames given and feeders repaired. The making of hive stands, bottoms, covers and other equipment as well as painting where needed, all help to shorten the winter months.

If we have any brood chambers partly filled with honey for spring feeding, we plan to clean these by scraping the propolis, cleaning any burr combs from between the frames and having them in shape for use. I like to give a hive body that has been cleaned as it most likely will be the one on which comb honey supers are placed the next summer. We must go through the brood chambers often and this extra clean-up makes manipulation easier in the busy season.

V. Care and Storage of Comb Honey

THERE has been considerable loss in comb honey in the past by failure of the producer to use wax moth fumigator or by using too small an amount to do a thorough job. The writer has seen beautiful comb honey in the grocery store that was badly damaged by moth.

One will not find moth, except in very rare cases, if a good job of fumigation is done in the shop before putting the sections into cellophane wrappers.

The supers, when brought into the shop, are placed in stacks of 15. This height is about the limit for lifting without too much effort. An empty super is placed on top, and inside this is placed the moth fumigating material which is carbon disulfide. The fumes of this liquid go downward during the evaporation process. It does not kill the eggs of the wax moth, but will kill the worm and the adult moth. Since it does not kill eggs it is necessary to fumigate the stacks of supers every seven or eight days as long as the stacks of supers remain n the shop, unscraped.

Carbon disulfide is highly inflammable and it must be used with care. It should not be used in any room where there may be danger of it exploding, or becoming ignited by sparks or an open flame.

Crystals such as paradichlorobenzene should never be used to fumigate comb honey. The honey will absorb the odor and it will not leave the honey. Sulfur candles are good if they can be used without danger of fire. They do not flavor the honey.

Fig. 48. All of these are a total loss.

Fig. 49. The top section of comb honey is free from fermentation. The bottom three sections show partial fermentation.

Removing Moisture From Comb Honey

The following is a reprint of an article by the writer, appearing in the January 1950 issue of the American Bee Journal.

It is hoped that we have at last found a favorable method to remove excess moisture from comb honey to prevent fermentation. For several years we have been working on this problem. Fermentation is not present every year in our honey. Some years the honey is of the very finest quality as far as moisture, flavor, and keeping qualities are concerned. It is only about once in five years that we have a loss from this source. Those who have not had any experience with fermentation cannot appreciate how fortunate they are. The results obtained during August and September research work have been most encouraging indeed. We have every reason to believe that we are finding the correct method to prevent future loss or to minimize it so that the loss will be light.

We produce section comb honey and the control of moisture in the comb has been quite a problem. If we were in liquid honey production the solution would be less troublesome.

Our greatest loss from fermentation has been in years of high humidity. We are not saying that high humidity is the only factor involved, but it is one of the major factors. We have reasons to believe that soil, humidity, plants, yeasts, and the bees themselves are all factors. We must include one more and that is temperature, because fermentation ceases or becomes inactive at temperatures below fifty degrees F. From our first experience it looked as though soil was to blame, the next time plants, and later, high humidity. As the years passed, we had to include all of the above as causing fermentation.

In 1929 we saw fermentation for the first time. Our apiaries were along the Indiana-Illinois state line. The soil was part clay and loam, with a small amount of black soil within flight of the bees. Only one apiary was on black soil. Honey produced in this apiary that year did not ferment; the honey from all other apiaries did show some fermentation. From 1929 to 1937 we had some loss but it was always from apiaries located upon the lighter soils. In 1936 we did not find one cell of sour honey; in fact, it was the finest quality we have ever produced, 12.3 pounds per gallon.

In 1941 it made little difference whether apiaries were on sandy, clay, loam or black soils, the honey all fermented. That year we lost over 700 supers of fine comb honey which would have graded number one to fancy, mostly fancy grade. It was not pleasant to see streams of foamy sour honey going down our shop drain. We used one apiary to clean out the supers storing the thin "soup" in deep frames for winter feed. The bees did a fine job on the second attempt.

For the past few years we have been trying to re-move moisture from the room where our filled comb supers were stored until the honey was ready for market. We believed that if we could keep the moisture in the room to a minimum, it might help draw moisture from the comb. We started using chemical units where the chemi-cals are suspended in a bag over a tank or pan. As mois-ture condenses, it falls into the pan and can be removed. The use of these units involved considerable work and the chemicals failed to act after the humidity dropped to a certain degree, unless the room was heated. These units did some good, we are sure, until we reached the season 1949, with its record of high humidity.

We practice the removal of comb supers from the hives as fast as the sections are completely sealed, to prevent travel stain and eliminate extra handling of equipment each time the colony is examined. Please do not tell us we should have left the supers on the hives longer to insure complete ripening of the honey. We have made tests and proved to ourselves that more honey was lost in the supers left on the hives than in what we removed. There appeared to be more moisture in the air than in the nectar the bees were trying to evaporate.

Early in 1949 we learned of a manufacturer who made machines for moisture removal. We exchanged a few letters but did not get one of their machines. We were sure, however, the machine was the answer to our prayer. The manufacturer was in another state and we hesitated about asking for a free demonstration or renting one for experimental use. If we had used this machine it would have saved us approximately 150 nice supers of comb. Later in the season we found another machine near us that was built to remove moisture. This machine is made by the Carrier Corporation who also made air-conditioning units. It is called the HUMIDRY and will withdraw five times as much water from the air as the chemical dehumidifier. It acts like a refrigerator running in reverse. The local distributor for this machine was the Punzak Air-Conditioning and Sales Company, Springfield Illinois. When I gave Mr. Punzak the history of our honey fermentation he was very much interested and suggested we use one of the machines for our experimental work.

Fig.50. This compact unit removes moisture from the comb honey.

Fig. 51. Showing the inside unit which acts like a refrigerator running in reverse.

The Humidry was placed in our comb room and turned on August 21 at 4:30 p.m. The outside temperature was about 85 degrees and humidity 66 per cent. There were 130 supers in the room at the time, also the chemical units which had been there for several days. These units were removed when the Humidry was turned on. A sample of honey was removed from a section to take a water content; it showed 21.0 per cent (sample A). On September 1, sample B was taken and showed 18.6 per cent; sample C taken on September 13, showed only 17.1 per cent. Here was the proof! We had removed moisture from the comb! Temperature and humidity readings were recorded twice daily during our test, water was weighted daily. From 4:30 p.m. August 21 to 8 p.m. September 13 we removed 222 ½ pounds of water from the Humidry. During this period the average temperature of the room was 79 plus F. and humidity 32 minus per cent.

After using the Humidry another season we should have a better report to make for we know this machine has a permanent place in our moisture control program. It should find a place with producers of cut comb and chunk comb, and even with the extracted honey producer who wants quality instead of quantity. We think we originated the id that if excess moisture is removed from any honey the flavor is improved; if we did not originate it, we firmly believe in it.

Date	Time	Humid.	Temp °F	Water Removed
Aug 21	4:30 pm	66	65	(Started today)
Aug 22	8:00 am	52	71	12 #
	8:00 pm	38	77	
Aug 23	8:00 am	33	84	13½ # (started
	8:00 pm	30	79	fan)
Aug 24	8:00 am	30	81	13 # (cut off
	8:00 pm	30	80	extra fan)
Aug 25	8:00 am	30	82	14 # (door
	8:00 pm	30	82	opened for 2hrs)
Aug 26	8:00 am	28	85	14 # (Humidry
	8:00 pm	28	84	off til 8am 27th)
Aug 27	8:00 am	48	75	5 ½ #
	8:00 pm	40	80	
Aug 28	8:00 am	40	80	17 #
	8:00 pm	38	80	
Aug 29	8:00 am	34	80	19 ½ #
	8:00 pm	34	80	
Aug 30	8:00 am	35	80	11 ½ #
	8:00 pm	35	80	
Aug 31	8:00 am	30	80	12 #
	8:00 pm	33	80	
Sep 1	8:00 am	33	78	7#
	8:00 pm			
Sep 2	8:00 am	31	75	7½ # (Humidry
	8:00 pm	32	72	off)
Sep 3	8:00 am	34	73	(started Hu-
	8:00 pm	38	72	midry at 8pm)
Sep 4	8:00 am	32	78	
	8:00 pm	32	80	

Sep 5	8:00 am	34	80	21 #
	8:00 pm	34	80	
Sep 6	8:00 am	33	78	5 #
	8:00 pm	32	78	
Sep 7	8:00 am	32	78	
	8:00 pm	32	77	
Sep 8	8:00 am	32	76	17 #
	8:00 pm	32	78	
Sep 9	8:00 am	28	64	8 #
	8:00 pm	29	74	
Sep10	8:00 am	28	72	
	8:00 pm	28	74	
Sep 11	8:00 am	30	72	6 #
	8:00 pm	32	72	
Sep 12	8:00 am	32	74	7 ½ #
	8:00 pm	36	74	
Sep 13	8:00 am	38	73	12 #
	8:00 pm	30	74	
Totals	**23 days**	**32-av.**	**79+av**	**222½ #**

Honey sample A 21.0 moisture (Aug 21)
Honey sample B 18.6 moisture (Sep 1)
Honey sample C 17.1 moisture (Sep 13)

VI. Marketing Comb Honey

TO merchandise any article successfully we must have a thorough knowledge of the goods to be sold. We must have a love of the product. We must first sell ourselves on the item. We must have a feeling that there is some good reason why we wish to sell the goods. It is better to leave honey unsold than to use some of the high pressure sales talk we hear so often.

There are three people to be satisfied when we market our honey. The first consideration should be the consumer who is to use it; second the dealer who has the distribution problem through the various outlets; and the third, we who produced the article. If the consumer is interested in our product, it naturally opens the way to benefit the dealer and producer.

Preparation for Market

A good worktable should be in every beekeeper's workshop. It requires considerable table space to do the necessary work required in removing the sections from the super, scraping off propolis and paraffin, grading, weighing, stamping, and packing. The table should be made fairly strong as several supers of honey may be on it at one time. It is best to have doors and drawers built into the underneath part of the table for the many things a person will want to keep in a table or cabinet of this kind. The table the writer uses (fig. 52) was built along these plans. The frame is 2x4 and the top is double, with oak flooring on the top. This oak covering gives the table a nice appearance and it is easily cleaned and receives an occasional polish. The drawers in the middle of the table contain such things as a hive tool, scraping knives, rubber stamps for net weighs, table brush, rubber thumb guards for scraping, razor blades, writing pad, and a host of other items. The doors at each end will give storage space for cellophane wrappers, paste, twine, paper tape, etc.

Fig. 52. A good worktable is necessary for handling comb honey. Doors and drawers beneath for storage.

The method used in the removal of honey from the ventilated T supers differs from that used in removing honey from supers having section holders. The super is placed upon the table right side up as in Fig. 53 and all four super springs are removed with a hive tool. The paraffin and any propolis on the sections is now scraped off. The best tool for this purpose is a single edge safety razor blade. (Fig. 54) May it be suggested that only new blades be used for sanitary reasons. The super is brushed off on top (Fig. 55) and turned bottom side up (Fig. 56), carefully using the right hand and arm to prevent the sections from falling out as the super is being turned. This is just an added precaution as any propolis, beeswax, or paraffin will hold the sections in the super. By placing both hands on the sections (Fig. 57) and applying a little pressure, the group of sections can be loosened and the super shell lifted from around the sections (Fig. 58). The T tins may now be removed with a slight upward stroke with the hive tool (Fig. 59).

Fig. 53. Removing 4 super springs with a hive tool.

After the removal of the super shell, this entire group of sections may be handled as one unit, as they should stick together (Fig. 50). In handling this group of sections one can place their forearm across the ends of the group and with a slight pressure with the other hand on the opposite end, the turning is made. All the turning is end over end. The bottoms are now scraped clean of paraffin and propolis, the same as the tops were. The entire group of 24 sections are now turned right side up again and each section removed and scraped individually (Fig. 61). Care must be taken as it does not take much of a scrape with the knife point to break the cappings and start a section to leaking.

In picking up a section, it is held in the left hand by the top and bottom, turning the section as it is scraped. In turning the section as it is scraped. In turning the section to reach all edges and corners with the knife, the thumb and fingers are placed on the top and bottom only, as the sections have been paraffined on these two surfaces and will not show any finger marks in handling (Fig. 62). Finger marks may show if sections are handled on unwaxed surfaces.

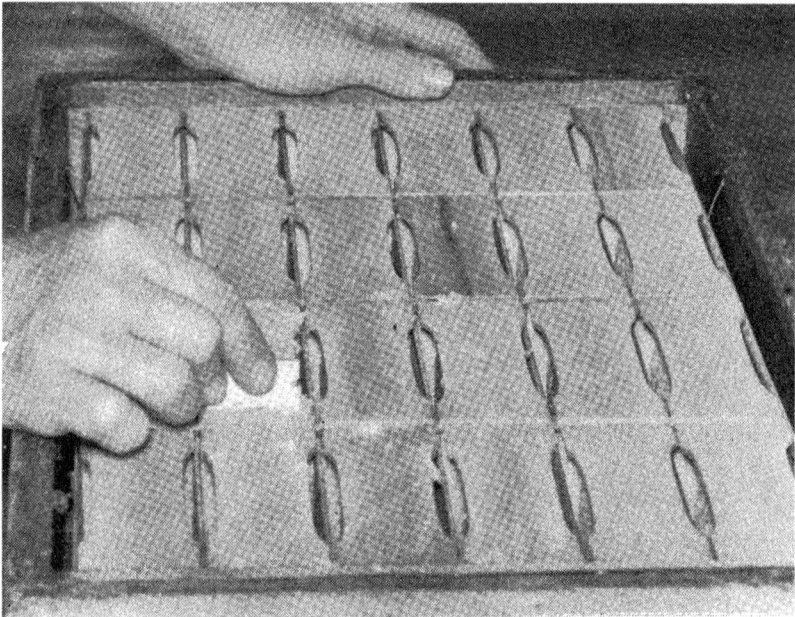

Fig. 54. The paraffin and any propolis is scraped off with a razor blade.

A homemade scraping knife is used as we do not find such a knife listed anywhere. Since the knife must be very sharp to do the work, it is best to protect the thumb with adhesive tape or use rubber thumb guards which are sold in stores handling office supplies.

As sections are scraped, the finished ones are placed in a neat stack so they will not become mixed when grading. The bait section with the "X" is left with the rest of the sections until graded, as the colony number is sometimes placed on it.

Grading and Weighing Comb Honey

In most cases all the sections in a super will pass into one grade. It is not very often that the sections in a super have to be sorted and placed into more than one grade. We find this to be the rule rather than an exception when using the ventilated T super.

The ungraded comb is stacked in neat stacks on the scraping table until time for grading. As each section is picked from the stack it is carefully examined on both sides to determine its grade as it is placed on the scales for weighing. The section is checked for completeness and whiteness of cappings, attachment of comb to wood, freedom from travel stain, peep holes or any damage that would not let it pass into a specified grade. At times one will find a section damaged where the comb had been attached to the separator with a small burr comb, or the comb slightly damaged with the scraping knife. Any section showing such damage is either placed in a lower grade or used for chunk comb.

Fig. 55. Brushing the top of the super clean.

Fig. 56. The super is turned over while carefully holding the sections with the right hand.

Fig. 57. By pressing with both hands, the sections are
loosened from the super shelf

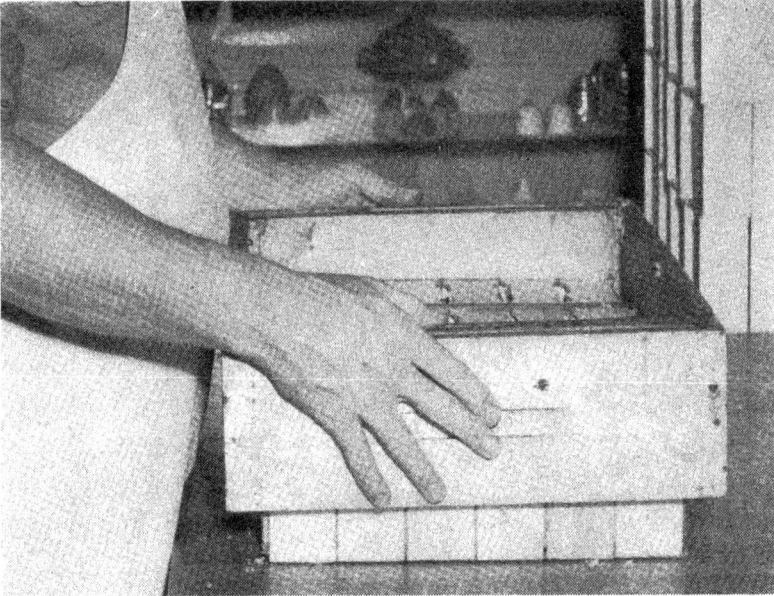

Fig. 58. The super shell is removed from the sections.

Fig. 59. The T tins are now removed with an upward stroke of the hive tool.

Fig. 60. Carefully hold the sections in this manner when turning the entire group end for end in cleaning.

Fig. 61. The individual sections now may be separated for further cleaning.

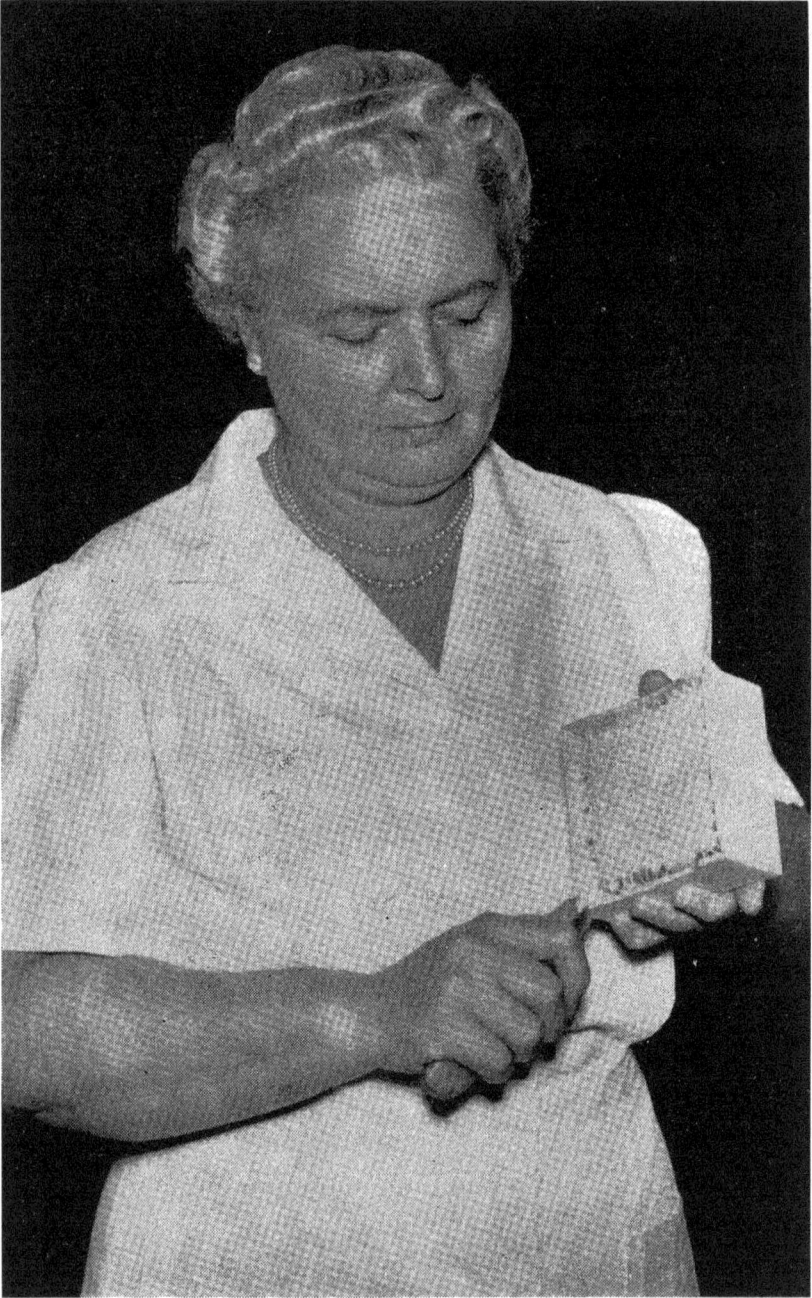

Fig. 62. Mrs. Killion, who prepares most of the crop for market, is holding a section by top and bottom.

Fig. 63.   Sections on the scale for weighing and stamping.

Fig. 64.   Wrapped comb honey in the shipping case.

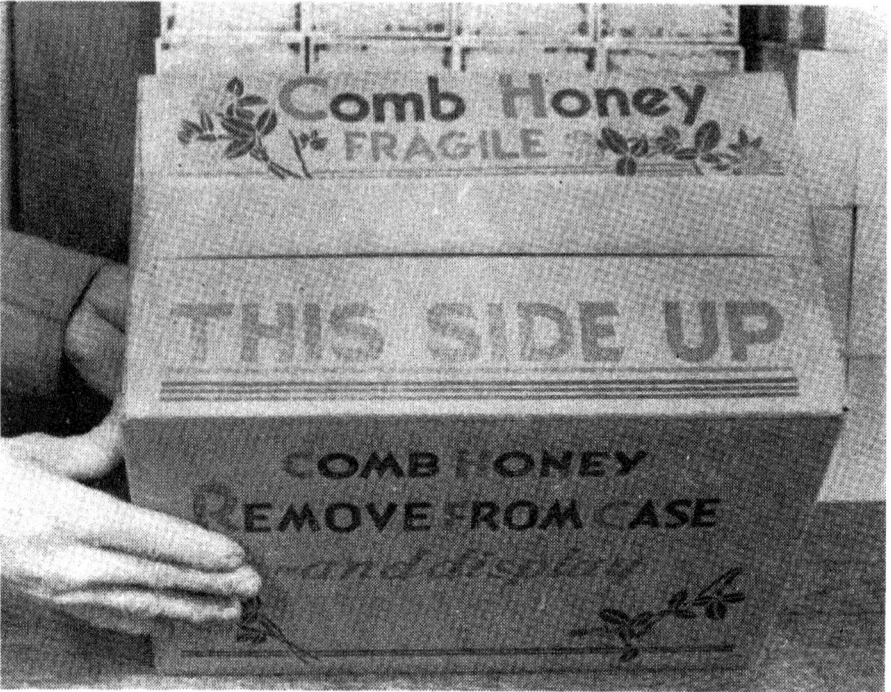

Fig. 65. The double-tier printed carton, holding 24 sections, is the most common type of shipping case.

Our grading rules differ somewhat from the present Federal grading rules. It is good to say that our rules have been slightly above those of the Federal rules in each grade. We expect a case of 24 sections to weigh at least 22 pounds and therefore are stamped 13 ounces net before it is placed in the Fancy Grade. This is an ounce above the Federal rules.

It has been suggested that our present grading rules should be changed to be more in line with the rules governing other agricultural products. Honey grades still refer to Extra Fancy, Fancy, Number One, and so on. The other agricultural products are graded as A, B, C, and so on. The writer would be favorable to such a change at any time as it appears necessary to make it.

If and when our rules are changed in any way, the writer is suggesting that our minimum net weight requirements be increased one ounce in each grade. Such an increase would not work a hardship on comb producers who wish only the finest quality to be marketed. This would improve the quality that would be offered for sale. At the present time there are not too many engaged in comb honey production and as the numbers increase they would be governed by the existing grading rules.

A color chart would perhaps be most helpful in a comparison of grades. The old color chart which was available a few years ago is now out of print and a new one will have to be made.

Some beekeepers weigh each section of honey on a small postal scale and stamp it individually. By this method one may have more than one group of weights in each case, unless the sections were stored and packed in various weight groups. We prefer to weigh an entire super of 24 sections at one time, providing they run fairly uniform in appearance. In an average season we do not find enough weight variation in this method to warrant the need for weighing each section separately.

A small set of platform scales is ideal for weighing comb honey. There are several good makes on the market and a person may select the one that fills the individual requirements best. In almost every beekeeper's shop there are a number of uses for a good set of scales. The platform should be large enough so that at least 12 sections may be placed on it at one time, and the next 12 can be placed on top of these. Some may want a platform that will take all 24 sections in a single tier. Such a large platform would take up more room however. We find our scale works very nicely; it holds 12 sections in each tier.

The tops and bottoms of the sections were painted with paraffin when the super was filled. Neither of these surfaces will do to stamp a net weight and name upon, therefore the sections must be turned so they may be stamped on the side. When placing the sections on the scales, they are turned on the side (or should we say edge).

We like to mark the net weight of the sections on the end of each case of honey. The package law in some states requires this marking.

No producer should ever be guilty of trying to hide some off grade honey in the bottom part of the case with a better grade on top. Each case should be as uniform as possible throughout. If any section falls below the others I grade, it should be removed and placed in the grade where it belongs.

Some beekeepers prefer to use the individual carton instead of the cellophane wrapper. They claim the carton affords a little more protection to the section than the cellophane will give. The carton will also hide more of the surface of the comb and has given an opportunity to pack an inferior quality of honey. The customer would have to open the carton to discover what quality of honey was on the inside. Most of these cartons have a rather small window through which to see the contents. On the other hand if the producer would pack an extra fancy grade in the carton he would hardly be doing the honey justice, especially in the ones with smaller windows.

Another type of wrapper which the writer does not approve is one which has cellophane on one side and paper covering the remainder of the section. This wrapper does not give the customer as good a chance to examine the unseen portion as easily as the carton.

It should be the desire of every beekeeper to show all the beauty of the comb; good quality comb honey does not deserve to be hidden.

There are a number of all cellophane wrappers on the market which do not hide the beauty of the comb although they do have printed borders. At one time an all cellophane printed bag wrapper was available, with a machine for aiding in the wrapping. This bag made a beautiful pack and sections could be wrapped in much less time than when the flat wrapper was used.

The writer prefers an all cellophane wrapper, printed with just enough design and wording to add color to the package. The wrapper, if made correctly, should not detract from the beauty of the honey or cover any defects. The printing on the wrapper is necessary to insure it being right side up at all times and shows the name and address of the producer or packer.

If the dovetails of the sections were all placed down as suggested in an earlier chapter, it will make a much neater package. If the dovetails are on the top side in the super they must remain that way in the wrapper.

It must be remembered too, that section comb honey is a seasonal article. It should be delivered to market in warm weather as it does not ship well in cool or cold weather, and this damage will make the honey run out and smear up the entire surroundings. Comb honey should be consumed before it has a chance to granulate. Once comb honey has granulated, the market for such is so limited as to make it almost worthless. About the only remedy is to melt it for liquid honey.

Marketing

There are several ways in which one may market comb honey. The plan followed in one locality may not work at all in a nearby area. We cannot say, "Here is the way to do it." The marketing of comb honey is not much different from that of other food products. The marketing picture never remains exactly the same; it is forever on the move, every changing, getting more complicated and interesting each year. Eating habits are different that they were a few years ago. The retail markets have undergone some radical changes, too. Many of the old neighborhood grocery stores have disappeared and in their place one finds the large super market with the seeing-eye door that swings open and shut without a touch. The market is brilliantly lighted inside with fluorescent lights that make the modern food packages sparkle like so many diamonds. Long, low shelves which are filled with expertly packaged food products, each with its individual eye appeal, await the customer in the markets of this kind.

In a room to the rear of all this, one can see into the packaging room. Here are men and women in white uniforms preparing meats, dairy products, dried fruits, and other materials to suit any discriminating customer. Our honey, therefore, must be in direct competition with all the other food products, dressed in their best. This is no place for a jar of honey with a smeared label partially granulated, or an unscraped section of comb with torn cellophane. Here is where quality of merchandise and intelligent packaging will pay good dividends. The honey must have what is known as customer or eye appeal, whether we are selling section comb, chunk comb, or extracted honey.

The neat roadside stand featuring honey, fruits, and nuts has always appealed to the writer as an ideal market, but there are very few such stands to be found in the Central States. We may be overlooking an excellent market. We have placed our honey in such a market and were pleased with the sales.

One feature which has been neglected in the marketing of honey is having trained demonstrators in large food markets. A trip through the largest stores in the Chicago Loop will prove the value of this type of sales promotion.

VII. Bulk Comb Honey

BEEKEEPERS will find the production of bulk comb honey interesting and profitable. The production of this type of honey will require more skill and closer observation of colonies than in extracted honey, but will not be as difficult to produce as section comb honey. Bulk comb honey is produced in shallow frames each holding 3 ½ to 4 ½ pounds depending upon the size and style of frame used. The honey in these frames may be cut into various sizes, drained and wrapped in cellophane and sold as cut comb honey. The pieces of comb may be placed in glass jars or pails, the remaining space filled with liquid honey and sold as chunk comb honey.

One may follow the same general plan in management of bulk comb colonies as he would for section comb. It is best in either case to use the stronger colonies for this type of honey. The only great difference in management is that section comb should always be produced over a single brood chamber, while bulk may be produced over a double brood chamber. The latter method would help reduce the colonies' desire to swarm. It is desirable to have the best combs of honey possible for either cut comb or chunk comb packages. The beauty of the comb will increase the sales, whereas faulty combs would detract and reduce sales.

Every section comb honey producer will find it advantageous to pack some chunk comb honey. He will find a certain number of people prefer this type of honey in preference to either comb or extracted. One customer said that chunk honey reminded him of "wild honey";

because he was such a consistent buyer we did not try to argue that there is no "tame honey" either.

We found that by packing some chunk honey each season we could use our lower grade of sections for this purpose instead of selling them as section comb honey. To sell this poorer quality would mean a lower price to the merchant and would tend to lower the price of better grades of comb honey. The comb in these lower grades is trimmed to make them suitable for the glass package and the sale brings a better price than if they had been marketed as section comb honey. When packed in glass it is considered a fancy pack of chunk honey. This package, therefore, does not in any way have a chance to lower prices. One should use only the sections containing light honey and when cut should contain only completely capped cells.

In the preparation of cut comb honey, the comb is cut to the desired size, and thoroughly drained before wrapping. It is important to drain every drop of liquid honey from the cut edges before wrapping as this loose honey will soon crystallize and the package will become unsalable. Any liquid honey will seep out of the wrappers if cartons are turned bottom side up. Cut comb honey should be marketed in such a way as to have immediate sale; it should never remain on the counter or shelf for a very long period of time. Some beekeepers are packing cut comb in various types of plastic containers which are ideal for advertising displays and in honey shows at fairs.

It requires less equipment for producing bulk honey than section comb, and less processing equipment than for extracted honey. One may produce more bulk comb in shallow frames per colony than comb honey in sections; but the price per pound will be less. There are two outlets in the sale of bulk comb honey, one is to sell the crop to a large packer or through a cooperative market, and the other is for the producer to pack and sell it. If the first plan is followed, it will not be necessary to bother with all the work of packing. If the producer wishes to pack his bulk comb in jars or pails as chunk comb honey it will necessitate the purchase of necessary equipment for this purpose. If the producer does not expect to bottle extracted honey except on a small scale and use the equipment for chink honey, he will not have to have too large an outfit.

Tanks and pipes or tubing used for heating and bottling should be of metal that will not injure the honey. Many producers are now having their tanks made of glass-lined steel, tinned copper, or stainless steel. The chemists tell us we should never use galvanized metal for our honey. Quite a few prefer stainless steel on account of its strength and general all-round use.

Before the purchase of frames and supers for bulk comb, it is best to decide what size or length of cut comb is to be desired. If the short 2 ½ pound square jar is to be used it will be best to produce a very shallow width comb. If the deeper jar like a 5-pound size is used, the combs should be deeper such as the 5 3/8 width frame. It is poor economy to produce a wide comb for a short jar and have all the wasted pieces left after trimming to size. It is also poor economy to use narrow starters or even half sheets of foundation in the frames, only full sheets should be used. When a narrow starter is used the bees most generally will build the remainder with drone cell size comb, and this is not as desirable as worker size.

Fig. 66. The 1 and 1 ½ pound sizes of chunk honey.

Fig. 67. The super may be used for the production of
shallow-frame honey. A closed-end frame is ideal for
shipping.

Fig. 68. A comb honey super converted into a shallow frame super for producing chunk honey.

Fig. 69. Most combs produced in the air-conditioned super are very uniform in appearance.

Fig. 70. Pan and drain board used when cutting the pieces of comb to proper size.

It has been found that melted beeswax is perhaps the best method for fastening the foundation in the frames.

Form boards, similar to the multiple block foundation fastening board, are convenient for fastening the foundation into frames. The board is cut to fit the inside of the empty frame, and slightly less than half the thickness of the frame. The reason for this thickness is to permit the foundation to be exactly center of then frame, when the sheet of foundation is laid upon the board.

Fig. 71. A one-pound jar of chunk honey.

Four or even six of these form boards may be fastened to a reel-like arrangement to speed up the fastening. Some kind of cleat or spring arrangement will have to be used to hold the frames in position while the reel is being turned. The beeswax should not be too dark, and it is best to melt it in a container set in hot water, instead of directly on the stove. A spoon may be used to pour the melted wax along the top edge of the foundation to fasten it to the top bar. This spoon should be bent at the tip to make it narrow and more convenient for pouring the hot wax where it is needed.

The value of the reel is demonstrated when pouring the wax in the top bar groove. The reel is turned so that one end of the frame is slightly higher than the other. Starting at the highest end of the top bar, a spoonful of wax is poured into the groove. This liquid wax is allowed to run to the lower end of the frame when the reel is turned to make the frame level. The sheet of wax is then pushed firmly into the melted wax and allowed to cool slightly before turning the reel. It will not take long before a person can time the work so he can make a perfect job at each attempt. It is always best to correct any sheets that are not securely fastened before they are given to the bees. One loose sheet falling part of the way out may cause the bees to fasten it to the frames on each side of it thereby damaging three frames instead of one.

We use a frame covered with ¼-inch mesh hardware cloth for cutting the comb into small sizes; this frame is used over a 2" deep pan we call the drip pan. We have pans about 30 inches square as this size suits our needs best. The pans should be made of heavy dairy tin with a ledge around the top to hold the frame. The covering of hardware cloth permits the combs to drain into pans. If several pans are used, the combs will have time to drain and will be perfectly dry of any loose honey before they are placed in the jars. If combs are not drained and liquid honey is in the bottom of the jar, this will start crystallization immediately.

Some packers of chunk honey have had special knives made to cut the entire comb out of the frame at one cut, and another special knife to cut the comb into the desired widths. This knife has blades electrically heated or heated in hot water. We have been using good sharp paring knives heated in a pail of water. As our chunk honey sales are good, we expect to use special gadgets to help wherever they can be used to any advantage.

When ready to fill the jars, the pieces of comb are placed into the jars first. Due to the fact that we have an abundance of comb and very little extracted honey, we fill our jars as full of comb as possible. A full pack of comb, neatly arranged, has eye appeal and means more repeat sales. The liquid honey is first heated to 160 degrees F. and cooled to 130 degrees F. before putting into jars. The heating is to destroy the yeasts in the honey, and to prevent or retard granulation. The honey, must, however be cooled to 130 degrees to prevent it melting the comb already in the jar. The caps are placed on the jars as they are filled and the jars are moved where they can be cooled before placing in shipping cases.

Any beekeeper will profit by not packing too large a quantity at one time or more than can be sold within a few weeks' time. If a grocer buys more than he can sell within a short time some of it may crystallize and be unsalable. It is always best not to overload any buyer with chunk comb honey.

An outlet for chunk comb honey is to be respected and not abused. It offers an opportunity to sell the consumer a quality product at a reasonable profit. The package helps in the sale of liquid honey which needs some additional sales effort to help relieve our present market for honey of this type.

VIII. Beekeeping Records

EVERY beekeeper should keep some kind of record of his work, or the bees' work. Perhaps the most unusual record we have is improperly called our weather book. We have recorded almost everything in this book along with data on the weather. The book is sort of a history. It tells of rainfall, droughts, snows, starting of honeyflows and many more items of special interest. Some of the pages tell of the big crops, others the small ones and even the failures. Sometimes we are sad as we note where a loved one has passed on. History is always like that, we must have the cloudy and rainy weather along with the bright sunshine.

Sometimes we may want to make a comparison of the seasons, when the first super was removed, when the weather was the driest. We find it all here in our book. The book was started a long time ago, on September 28, 1910. It is now over forty years old and it tells of many changes. One page in the book is quite interesting. On June 23, 1927 (speaking of the honeyflow), it says, "seems as though it will fail." More remarks a few days later was that the flow as getting better and bees were pouring into the hives on certain days. On July 11 it read "supers are getting low." The supply of supers in the shop were about all used up. That year was the best year on record, and on June 23 it had seemed so likely that weather conditions would cause it to fail. That is only one of many interesting surprises we find by rechecking the record book from time to time.

It is interesting to know the average blooming dates for honey plants over a period of years. For central Illinois we find the average date for our clover flow to start is June 9. In a forty-year period our records show May 16 as the earliest and June 28 as the latest.

The year 1936 was the hottest and perhaps the driest ever recorded in East Central Illinois. The first half of July was extremely hot with temperatures well above 100 degrees F every day and 19 degrees on the 14th. The honey crop for that year was good and quality excellent. The clovers had received an adequate supply of moisture to produce nectar. When the dry weather and heat came, the clovers continued to produce nectar.

One fine thing in looking through this book is the courage it gives one to continue on in the bee business even when everything appears to be going against you. We read of the bad seasons and how we pulled through, and along came a good year and everything was in fine shape. There has never been a real failure any year in beekeeping. Some years (and recently) we didn't get any honey, and did not make expenses, but we learned a world of knowledge and that is profitable. It is the poor years that sometimes make us think more of and to respect the good seasons when they com.

Our old worn record book does a lot of good when we read over the interesting things that we wrote years ago.

IX. The Future for Comb Honey

THE writer has always been optimistic about the future for comb honey. He is even more optimistic now for the future of both section and bulk comb honey. Perhaps we will never again find so few beekeepers producing this quality product. The low point in comb production has been reached and we can expect a rapid increase in production. Let us hope the time will never come again when so many who prefer comb honey will have to depend upon so few of us for it. In the past two years the writer has traveled many thousands of miles, into several states, to encourage beekeepers to produce more comb honey. It now appears that the trips were not in vain.

There has been considerable work done in the field of plastic containers for cut comb honey and some of these containers really have merit. They will no doubt, help market a considerable amount of this kind of honey, as the idea is new and will be used by some buyers who may never have handled honey before.

We have been thinking and working along the idea of plastics and other new materials for several years, and have been trying to see just where the use of plastics would fit into the picture. We have debated as to the probability of plastics replacing the wood section. What would happen if we were to try various colored plastic sections? It is our opinion that honey in the comb appears out of place when dressed in plastic. Plastics and other materials are too modern and have too much of that mass production look to associate them with comb honey. This honey must have that individual appearance, that the bees made it just for you. We have never found anything that would cause us to want to change from the idea of the wood honey section. Maybe we are a little old-fashioned in our way of thinking. Sometimes that is the best way. The bees have never changed their way of making honeycomb all through the ages. They are doing it the same today as they did centuries ago. There is something inseparable, we think, when we refer to honey-comb—its attachment to wood. There is something about this that is historical and beautiful. Honey in the comb. It was the first sweet known to man.

"My son, eat thou honey, because it is good; and honeycomb which is sweet to thy taste: so shall the knowledge of wisdom be unto the soul."

Honey in the comb, in a wooden frame; surely this is one of the world's most beautiful and unusual pictures.

THE END

Index

About the Author

Carl E. Killion Jr. at the Ohio State Beekeepers Association meeting in Defiance, Ohio in 1967.

Carl Everest Killion was born September 2nd 1899 in a log cabin several miles from the community of Diamond, Indiana. He became interested in bees and covered bridges at an early age recalling his first taste of honey at age six and weathering a severe thunderstorm under the protection of a newly built covered bridge. He was reared through the great depression and due to the times was unable to obtain a high school education. His father sustained their family employed in the coal mines, where Carl himself would later work.

Carl's 1966 book "The Covered Bridge" relayed some of the key events of his child hood and how he came to have a fondness for covered bridges and bees.

Carl's father not enamored with bees and was not agreeable to spend hard earned dollars for beekeeping adventures. Carl's work allowed him access to wooden apple crates which he scavenged and formed into bee frames and boxes, but finding wild bee swarms proved much harder. His first set of bees came from a bee tree that he and his friends were able to acquire and by season's end Carl had established fifteen hives. His family later moved to Paris, Illinois, where he continued his interest in bees and beekeeping. He became a founding member of the state's beekeeping society and later the state's apiary inspector and bee expert.

He met and married Elizabeth Hayes and began a family, but unable to make an adequate living at that time with just beekeeping he stayed with mining until 1935 when a mine accident caused him to re-think mine labor.

His interest in bees began with observation of local beekeepers and their efforts in capturing wild bees and continued throughout the years discussing, collaborating, teaching and legislating with beekeeping colleagues. A 1951 summary of his own beekeeping experiences are found in this book, "Honey in the Comb", originally self-published by Killion and Son's Apiaries. He was a close friend of Charles Dadant.

Carl passed away June 1st 1979.

www.ingramcontent.com/pod-product-compliance
Lightning Source LLC
Chambersburg PA
CBHW031940190326
41519CB00007B/598